Advances in Multicriteria Analysis

Nonconvex Optimization and Its Applications

Volume 5

The titles published in this series are listed at the end of this volume.

Advances in Multicriteria Analysis

Edited by

Panos M. Pardalos
University of Florida

Yannis Siskos
Technical University of Crete

and

Constantin Zopounidis
Technical University of Crete

KLUWER ACADEMIC PUBLISHERS
DORDRECHT / BOSTON / LONDON

Library of Congress Cataloging-in-Publication Data

Advances in multicriteria analysis / edited by Panos M. Pardalos,
Yannis Siskos, Constantin Zopounidis.
p. cm. -- (Nonconvex optimization and its applications ; v.
5)
Includes index.
ISBN 0-7923-3671-2 (hb : alk. paper)
1. System analysis. 2. Decision-making--Mathematical models.
I. Pardalos, P. M. (Panos M.), 1954- . II. Siskos, Yannis.
III. Zopounidis, Constantin. IV. Series.
T57.6.A35 1996
658.4'03--dc20 95-34337

ISBN 0-7923-3671-2

Published by Kluwer Academic Publishers,
P.O. Box 17, 3300 AA Dordrecht, The Netherlands.

Kluwer Academic Publishers incorporates
the publishing programmes of
D. Reidel, Martinus Nijhoff, Dr W. Junk and MTP Press.

Sold and distributed in the U.S.A. and Canada
by Kluwer Academic Publishers,
101 Philip Drive, Norwell, MA 02061, U.S.A.

In all other countries, sold and distributed
by Kluwer Academic Publishers Group,
P.O. Box 322, 3300 AH Dordrecht, The Netherlands.

Printed on acid-free paper

Printed in the Netherlands

"The things of which there can be sight,

hearing, and learning these are what

I especially prize"

Heraclitus

CONTENTS

IV. MULTI-OBJECTIVE OPTIMIZATION

V. REAL-WORLD APPLICATIONS

EDITORIAL

A historical perspective

For the last 25 years a decision-aid approach called Multicriteria Analysis (M.A.) has been under development. This approach leads to the development of the classic optimization paradigm in operations research. The great success of M.A. today is manifested by the increasing number of theoretical and application papers which have been published in scientific journals of operations research and decision science and by the great number of communications presented in scientific meetings. Furthermore, several international working groups have been established to analyze this topic on a regular basis. M.A. is often called multiple criteria decision making (MCDM) by the American School and Multicriteria decision aid (MCDA) by the European School.

The history of M.A. starts with the work of Pareto (1896) where the problem of the aggregation of criteria into a single criterion is examined. Pareto also defines the concept of efficiency between two decision alternatives.

As its name indicates, M.A. aims to provide the decision maker with some tools in order to enable him to advance in solving a decision problem where several often conflicting points of view must be taken into consideration.

Actually, one can distinguish several categories of methods in M.A. The boundaries between these categories are, of course, rather fuzzy. Roy (1985) proposes the following three categories of methods:

(1) unique synthesis criterion approach disregarding any incomparability,

(2) outranking synthesis approach, accepting incomparability, and

(3) interactive local judgement approach with trial-error iterations.

According to Scharlig (1985), the above three categories are called respectively: complete, partial and local aggregation methods. Vincke (1992) calls them respectively: multiattribute utility theory (MAUT), outranking relation methods and interactive methods. Zionts (1992) divides the M.A. methods into four subareas: (1) multiple criteria mathematical programming, (2) multiple criteria discrete alternatives, (3) multiattribute utility theory, and (4) negotiation theory.

Finally, it is possible to propose a framework for dividing the M.A. into four categories:

(1) multiobjective mathematical programming,

(2) multiattribute utility theory,

(3) outranking relations approach, and

(4) preference disaggregation approach.

In multiobjective mathematical programming, Koopmans (1951), in his work on activity analysis of production and allocation, has developed the concept of efficient vector or non-dominated vector. Lately, this concept has been used by Markowitz (1952) in his work on portfolio selection.

A direct extension of Koopmans' ideas also appears in the original book of Charnes and Cooper (1961) on goal programming. Goal programming is one of the best known methods in M.A. It works with linear programming models and involves choosing targets or goals for each objective. Goal programming's development continued under the leadership work of Ijiri (1965), Lee (1972) and Ignizio (1976).

Based on the work of Koopmans, vector optimization techniques for computing the set of all non-dominated solutions of a multiobjective program have been developed by many authors (Geoffrion, 1968; Philip, 1972; Evans and Steuer, 1973; Yu and Zeleny, 1975; Gal, 1977; Isermann, 1977; Winkels, 1980; Ecker et al., 1980; Goicoechea et al., 1982). An excellent overview of these techniques has been made by Gal (1982).

Because the size of the non-dominated set can make it difficult to choose a final solution, interactive procedures have been developed since 1970(Benayoun et al.,

1971; Geoffrion et al., 1972; Dyer, 1972; Zionts and Wallenius, 1976; Wierzbicki, 1980; Franz and Lee, 1981; Spronk, 1981; Steuer and Choo, 1983; Nakayama and Sawaragi, 1984; Korhonen and Laakso, 1986; Teghem et al., 1986; Korhonen and Wallenius, 1988; Steuer et al., 1993).

In MAUT, the decision-aid problem consists of modelling the decision maker's preferences by means of a value function (decision under certainty) or a utility function (decision under uncertainty) which is, mainly, supposed to be additive. The axiomatization of the additive utility model was made by Debreu (1960), Luce and Tukey (1964), Krantz (1964), Krantz et al. (1971) and Wakker (1989). In the uncertainty case, Aumann (1964) seems to be the first to adapt the expected utility theory of Von Neumann-Morgenstern (1944) in a multicriteria context. The necessary conditions for multiattribute models in expected utility theory are provided by Pollak (1967), Keeney (1968), Raiffa (1969), Fishburn (1970) and others. An excellent review of MAUT models appears in Keeney and Raiffa (1976) and Dyer and Sarin (1979).

It is possible to include in MAUT the Analytic Hierarchy Process (AHP) of Saaty (1980). The method is embodied in software known as Expert Choice (Forman et al., 1983).

The concept of outranking is attributed to Bernard Roy, who can be considered the founder of these methods. This concept was born out of difficulties encountered with diverse concrete problems.

Roy (1968) presents ELECTRE I (ELimination Et Coix Traduisant la REalite), which is based on a relational representation of the decision maker's preferences. According to Roy (1974), an outranking relation is a binary relation S defined in a set of alternatives A such that alternative a outranks alternative b (aSb) if, given what is known about the decision maker's preferences and given the quality of the evaluations of the alternatives and the nature of the decision problem, there are sufficient arguments to decide that a is at least as good as b, while there is no essential reason to refute this statement.

ELECTRE II, III and IV are developed and presented respectively by Roy and Bertier (1973), Roy (1978), and Roy and Hugonnard (1982). Roy and Skalka (1984) present the software ELECTRE IS which is inspired directly from ELECTRE I. Yu (1992) presents ELECTRE TRI which is a development of ideas of Moscarola and Roy (1977) and is devoted to the sorting problems.

Other outranking relation methods are QUALIFLEX (Paelinck, 1978), ORESTE (Roubens, 1981), MELCHIOR (Leclercq, 1984) and PROMETHEE (Brans et al., 1984; Brans and Vincke, 1985). In the uncertainty case, outranking relation methods were developed by Jacquet-Lagrèze (1975), Martel and D'Avignon (1982), Siskos (1983) and D'Avignon and Vincke (1988). In the works of Siskos et al. (1984) and Roy and Bouyssoy (1993), one can find a great number of references related to applications of these methods.

Disaggregation analysis is often used in M.A. as a means for modelling the preferences of an individual or a group of decision makers. This analysis tries to find what criteria aggregation model is behind an overall preference result (pairwise comparisons, ranking, etc). Regression-based approaches have been developed and are proposed as an alternative to MAUT for utility assessment in M.A. Both approaches generally refer to a common axiomatic basis that often assumes common utility-decomposition forms, and they differ mainly in the methodology for assessing the utility models. In MAUT, the parameters of the utility-decomposition model are estimated in a direct way on the basis of some preference information obtained from the decision maker in terms of judgements on criteria values. According to the disaggregation approach, these parameters are estimated by analyzing the decision maker's overall preference on some reference alternative. The problem is then to estimate an additive utility function that is as consistent as possible with the known subjective preferences of a decision maker.

Regression-based methods were proposed by Srinivasan and Shocker (1973), Pekelman and Sen (1974), Young et al. (1976), Jacquet-Lagrèze and Siskos (1982) and Oral and Kettani (1989). Jacquet-Lagrèze and Siskos developed the UTA (UTilite Additive) method which consists of first determining an optimal utility function through linear programming techniques and then performing a sensitivity analysis in order to construct a set of utility functions, all of them being models consistent with the decision maker's a priori preferences. Some improvements of the UTA method were made by Siskos and Yannacopoulos (1985) and Despotis et al. (1990). Furthermore, the UTA algorithm was used as a learning procedure of the decision maker's preferences in the works by Stewart (1987) and in the multicriteria decision aid systems PREFCALC (Jacquet-Lagrèze, 1984) and MINORA (Siskos et al., 1993).

Organizational aspects

The field of M.A. is currently well organized. According to a study by Steuer et al. (1994) which presents recent statistical data, there are three international and four national societies.

The three international societies are:

(1) International Society on Multiple Criteria Decision Making (1100 members from over 70 countries, formed in 1979);

(2) EURO Working Group on Multicriteria Decision Aid (300 members from over 30 countries, formed in 1975); and

(3) European Summer Institute Group on Multicriteria Analysis (200 members from over 30 countries, started in 1985).

The aim of the International Society on MCDM is to facilitate worldwide communications among all people interested in M.A. For this purpose, the Society publishes the newsletter **MCDM WorldScan** (started in 1987) and distributes it to all its members. The EURO Working Group has the following objectives:

(1) to contribute to developing an original way of thinking, at a European level, in the field of multicriteria decision making;

(2) to allow each member of the group to present his research, submit those to critical discussion in the group, and to present open problems;

(3) to keep the continuity in the exchange of reflections from one meeting to another; and

(4) to maintain the environment of a working group and not to transform the meeting into a small congress.

ESIGMA's aim is to allow a more informal discussion of the ideas in M.A.

The national societies are:

(1) Chinese Special Interest Group in MCDM (started in the early 80's);

(2) Czechoslovak National Group on Multicriteria Optimization (started in the late 80's);

(3) German Working Group "Mehrkriterielle Entscheidung" (started in the former GDR Germany in 1976); and

(4) Polish Group on Polyoptimization in Design (started in the early 80's).

In addition, the International MCDM Summer School organizes two-week summer sessions for new researchers and other young professionals interested in M.A. Since 1983, there have been five such summer schools organized.

Up to the present, there have been eleven (11) conferences of the International Society on MCDM, forty (40) meetings of the EURO Working Group on MCDA and nine (9) meetings of ESIGMA (cf. Steuer et al., 1994, for more details).

In addition, the field of M.A. now has its own journal called **Journal of Multi-Criteria Decision Analysis** which started in 1992. It is published by Wiley.

Among the international refereed journals, the European Journal of Operational Research has published by far most of the articles of M.A. Following this journal are the Journal of the Operational Research Society, Computers and Operations Research, Mathematical and Computer Modelling, Journal of Optimization Theory and Applications, and so forth.

Future directions for research

After the great explosion of M.A. in the last twenty years, the question is how to predict the future direction of this science. Many researchers have considered this problem.

First, Bouyssou et al. (1993) in their work "The manifesto of the new MCDA era" present the MCDA as:

- an attitude toward providing decision-aid to actors involved in a decision process;
- a methodology for providing such a decision-aid;
- a collection of methods, and
- a corpus of experience obtained after many real world applications.

For these authors, M.A. is an open theoretical field and not a closed mathematical one, and they propose the following areas for future research:

- Key theoretical and methodological issues have to be addressed;
- New potential topics of applications have to be explored;
- "True" applications are needed as well as ex-post analysis; and
- Connections between M.A. and other areas of research have to be explored (i.e. artificial intelligence, control theory, industrial planning, logistics, etc.).

The need to have more applications and tests of the available multicriteria methods are found in the works of Dyer et al. (1992), Stewart (1992), and Pomerol and Barba-Romero (1993).

Another important future direction of the M.A. is the development of multicriteria decision support systems (MCDSSs). A MCDSS is simply a DSS that helps to implement MCDM methods. Characteristics of MCDSSs include analyses of multiple criteria, involvement of MCDM methods, and the integration of user input (decision maker's preferences) in the modelling processes (cf. Jelassi et al., 1985). The development of some MCDSSs can be found in Siskos and Zopounidis (1993) and Jelassi and Brans (1994). The importance of MCDSSs in the modelling process of the decision maker's preferences is cited in the works of Dyer et al. (1992), Korhonen et al. (1992), and Pomerol and Barba-Romero (1993).

The negotiation theory is regarded as a natural extension of MCDM (Zionts, 1992). Pruitt (1981) defines negotiation as a process by which a joint decision is made by two or more parties. As Roy (1994) remarks, the multicriteria decision aid can provide a substantial support in a group decision-making situation where there are considerable value conflicts between group members. The extension of MCDM methods in the negotiation aid is also proposed by Stewart (1992).

Roy (1994) proposes the following important questions about the future of M.A. in order to complete the propositions made by Bouyssou et al. (1993) in their manifesto:

- What is the scientific basis of a decision aid science?
- What is the relationship between the decision aid science and the science of decision behavior?
- Does one have a habit of confusing the result of a procedure with the recommendation?

In the last question, the objective is to help the decision maker construct a model of preferences on the basis of work assumptions which are not taken as norms. These work assumptions must be considered as recommendations rather than as norms (constructive approach, cf. Bana e Costa, 1993).

Finally, Dyer et al. (1992) (The American School of M.A.) propose the following areas for the next ten years of research (except the MCDSSs discussed above):

- behavioral aspects (the question is how one should incorporate the existing behavioral and psychological knowledge in designing multicriteria decision aid systems);
- robustness considerations (one introduces sensitivity analysis and vague or imprecise judgements and/or probabilities in multiattribute situations and decisions under uncertainty in which states are multidimensional); and
- the role of heuristics (the appeal of heuristic methods stems from their ability to quickly generate near-optimal solutions to difficult problems; see the AIM approach of Lotfi et al., 1990).

Outline of the book

The contents of the book are organized into five sections: Methodology, Problem structuring, Utility assessment, Multiobjective optimization and Real-world applications.

In the methodology section, using a small example, Vincke proposes a methodology for studying one of the points stressed in the "Manifesto of the new MCDA era" and which is concerned with the need for a basic theory of MCDA procedures. The paper by Bouyssou deals with the problem of choosing on the basis of a fuzzy preference relation. Its purpose is to characterize what appears to be a "reasonable" choice procedure in the particular case in which the meaning of fuzziness is strictly ordinal. Mousseau studies the notion of Relative Importance of Criteria (RIC) by Elicitation Techniques for Importance Parameters (ETIP). He proposes an ETIP adapted to the ELECTRE methods that proceeds by means of pairwise comparisons of fictitious alternatives.

In the problem structuring section, Larichev and Moshkovich study a class of unstructured decision making problems and present the method ZAPROS, based on ordinal trade-offs, as a decision aid tool to solve these problems. The purpose of the paper by Korotkich is to focus attention on the role of a structural complexity as a tool for the criteria definition and study of optimal problem solving.

In the utility assessment section, Bana e Costa and Vansnick describe the theoretical basis of the MACBETH method which is an interactive approach for cardinal measurement of judgements and preferences of a decision maker. Despotis and Zopounidis, in their paper, present the formulation of an additive model which treats both monotonic and non-monotonic partial utilities on a common basis and in that sense can be viewed as a generalization of the UTA ordinal regression method.

In the multi-objective optimization section, Zeleny proceeds to the study of the Ideal-Degradation Procedure (IDP), which is designed to solve multiple criteria problems without aggregation or scalarization. The paper of Jaszkiewicz and Slowinski presents a new interactive procedure for multiple objective analysis of linear and non-linear programs. The innovative element of this procedure is that the set of non-dominated points provided to the decision maker comes from an outranking relation. The new procedure is called the Light Beam Search and is totally computerized. The paper by Nakayama is a review of the satisficing trade-off method by presenting its theoretical foundations and some applications (i.e. the bond portfolio problem). Migdalas examines the conditions under which a Stackelberg equilibrium is, indeed, a Pareto optimum. For this purpose, he used the bi-criteria optimization problem.

In the real-world applications section, Siskos, Baourakis, Grigoroudis and Matsatsinis present a preference disaggregation approach in order to analyze the consumer behavior in agricultural management (i.e. new product development, identification of market shares, launching strategies, etc.). A real-world application of the UTA disaggregation method to evaluate R and D projects is found in the paper of Jacquet-Lagrèze. The paper by Costa and Climaco shows how a MOLP multiple reference point approach (MRPA) can proceed to the modelling of a financial plan over a planning period. Finally, Benson describes the nature and use of the linear multiple objective integer programming model for scheduling surgeries requiring post-operative intensive care.

This book represents a small contribution to the state of the art of M.A. We are grateful to the contributors who agreed to submit their work for publication in this volume. We are also grateful to the numerous referees who have genuinely evaluated the submitted papers and all those who have provided us with assistance in the material collection and management.

Panos M. Pardalos

Dept. of Industrial and Systems Engineering

303 Weil Hall

University of Florida

Gainesville, FL 32611, U.S.A.

Yannis Siskos and Constantin Zopounidis

DSS Laboratory

Dept. of Production Engineering and Management

Technical University of Crete

University Campus

73100 Chania, Greece

References

Aumann, R.J. (1964) Subjective programming, in: Shelby, M.W. and Bryan, G.L. (eds), *Human Judgements and Optimality*, John Wiley, New York, 217-242.

"Bana e Costa, C. (1993) Convictions et aide a la decision, *Newsletter of the European Working Group "Multicriteria Aid for Decisions"* 2, 3, 1-3.

Benayoun, R., J. de Montgolfier, J. Tergny and O. Larichev (1971) Linear programming with multiple objective functions: Step method (STEM), *Mathematical Programming* 1, 366-375.

Bouyssou, D., P. Perny, M. Pirlot, A. Tsoukias and Ph. Vincke (1993) A manifesto for the new MCDA era, *Journal of Multi-Criteria Decision Analysis* 2, 3, 125-127.

Brans, J. P., B. Mareschal and Ph. Vincke (1984) PROMETHEE: A new family of outranking methods in multicriteria analysis, in: Brans, J.P. (ed.), *Operational Research* 84, North-Holland, Amsterdam, 408-421.

Brans, J.P. and Ph. Vincke (1985) A preference ranking organization method, *Management Science* 31, 6, 647-656.

Charnes, A. and W.W. Cooper (1961) *Management models and industrial applications of linear programming,* vol. 1, John Wiley, New York.

D'Avignon, G.R. and Ph. Vincke (1988) An outranking method under uncertainty, *European Journal of Operational Research* 36, 311-321.

Debreu, G. (1960) Topological methods in cardinal utility theory, in: Arrow, K.J., Karlin, S. and Suppes, P. (eds), *Mathematical Methods in the Social Sciences,* Stanford University Press, Stanford, CA.

Despotis, D.K., D. Yannacopoulos and C. Zopounidis (1990) A review of the UTA multicriteria method and some improvements, *Foundations of Computing and Decision Sciences* 15, 2, 63-76.

Dyer, J.S. (1972) Interactive goal programming, *Management Science* 19, 1, 62-70.

Dyer, J.S. and R.K. Sarin (1979) Measurable multiattribute value functions, *Operations Research* 27, 810-822.

Dyer, J.S., P.C. Fishburn, R.E. Steuer, J. Wallenius and S. Zionts (1992) Multiple criteria decision making, Multiattribute utility theory: The next ten years, *Management Science* 38, 5, 645-654.

Ecker, J.G., N.S. Hegner and I.A. Kouada (1980) Generating all maximal efficient faces for multiple objective linear programs, *Journal of Optimization Theory and Applications* 30, 353-381.

Evans, J.P. and R.E. Steuer (1973) Generating efficient extreme points in linear multiple objective programming: Two algorithms and computing experience, in: Cochrane, J.L. and Zeleny, M. (eds), *Multiple Criteria Decision Making,* University of South Carolina Press, Columbia, SC, 349-365.

Fishburn, P.C. (1970) *Utility theory for decision making,* John Wiley, New York.

Forman, E.H., T.L. Saaty, M.N. Selly and R. Waldron (1983) *Expert Choice,* Decision Software, Inc., Pittsburgh.

Franz, L.S. and S.M. Lee (1981) A goal programming based interactive decision support system, in: Morse, J.N. (ed.), *Organizations: Multiple Agents with Multiple Criteria,* Springer-Verlag, Heidelberg, 110-115.

Gal, T. (1977) A general method for determining the set of all efficient solutions to a linear vector maximum problem, *European Journal of Operational Research* 1, 307-322.

Gal, T. (1982) On efficient sets in vector maximum problems: A brief survey, in: Hansen, P. (ed.), *Essays and Surveys in MCDM,* Springer-Verlag, New York.

Geoffrion, A.M. (1968) Proper efficiency and the theory of vector maximization, *Journal of Mathematical Analysis and Applications* 22, 618-630.

Geoffrion, A.M., J.S. Dyer and A. Feinberg (1972) An interactive approach for multicriterion optimization with an application to the operation of an academic department, *Management Science* 19, 357-368.

Goicoechea, A., D.R. Hansen and L. Duckstein (1982) *Multiobjective decision analysis with engineering and business applications,* John Wiley, New York.

Ignizio, J.P. (1976) *Goal programming and extensions,* D.C. Heath, Lexington, Massachusetts.

Ijiri, Y. (1965) *Management goals and accounting for control,* North-Holland, Amsterdam.

Isermann, H. (1977) The enumeration of the set of all efficient solutions for a linear multiple objective program, *Operations Research Quartely* 28, 711-725.

Jacquet-Lagrèze, E. (1975) La modelisation des preferences: Preordres, quasi-ordres et relations floues, *These de 3e Cycle,* Universite Paris V, Paris.

Jacquet-Lagrèze, E. (1984) PREFCALC: Evaluation et decision multicritere, *Revue de l'utilisateur de IBM-PC,* no 3, 38-55.

Jacquet-Lagrèze, E. and J. Siskos (1982) Assessing a set of additive utility functions for multicriteria decision-making, the UTA method, *European Journal of Operational Research* 10, 151-164.

Jelassi, M.T. and J.P. Brans (eds) (1994) Special issue with selected papers from the IFORS-SPC 1 Conference on Decision Support Systems, *Decision Support Systems* 12, 249-250.

Jelassi, M.T., M. Jarke and E. Stohr (1985) Designing generalized multiple criteria decision support system, *Journal of Management Information Systems* 1, 24-43.

Keeney, R.L. (1968) Quasi-separable utility functions, *Naval Research Logistics* 15, 551-565.

Keeney, R.L. and H. Raiffa (1976) *Decisions with multiple objectives; Preferences and value trade-offs,* John Wiley and Sons, New York.

Koopmans, T.C. (1951) Analysis of productions as an efficient combination of activities, in: Koopmans, T.C. (ed.), *Activity Analysis of Production and Allocation,* Cowles Commission Monograph no 13, John Wiley & Sons Ltd., New York.

Korhonen, P.J. and J. Laakso (1986) Solving generalized goal programming problems using a visual interactive approach, *European Journal of Operations Research* 26, 355-363.

Korhonen, P.J. and J. Wallenius (1988) A Pareto race, *Naval Research Logistics* 35, 6, 615-623.

Korhonen, P.J., H. Moskowitz and J. Wallenius (1992) Multiple criteria decision support: A review, *European Journal of Operational Research* 63, 361-375.

Krantz, D.H. (1964) Conjoint measurement: The Luce-Tukey axiomatization and some extensions, *Journal of Mathematical Psychology* 1, 248-277.

Krantz, D.H., R.D. Luce, P. Suppes and A. Tversky (1971) *Foundations of measurement,* vol. I, Academic Press, New York.

Leclercq, J.P. (1984) Propositions d'extension de la notion de dominance en presence de relations d'ordre sur les pseudo-criteres: MELCHIOR, *Revue Belge de Recherche Operationnelle, de Statistique et d'Informatique* 24, 1, 32-46.

Lee, S.M. (1972) *Goal programming for decision analysis,* Auerbach Publishers, Philadelphia.

Lotfi, V., T.J. Stewart and S. Zionts (1992) An aspiration-level interactive model for multiple criteria decision making, *Computers and Operations Research* 19, 7, 671-681.

Luce, R.D. and J.W. Tukey (1964) Simultaneous conjoint measurement: A new type of fundamental measurement, *Journal of Mathematical Psychology* 1, 1-27.

Markowitz, H. (1952) Portfolio selection, *The Journal of Finance,* March, 77-91.

Martel, J.M. and G.R. D'Avignon (1982) Projects ordering with multicriteria analysis, *European Journal of Operational Rcsearch* 10, 56-69.

Moscarola, J. and B. Roy (1977) Procedure automatique d'examen de dossiers fondee sur une segmentation trichotomique en presence de criteres multiples, *RAIRO Recherche Operationnelle* 11, 2, 145-173.

Nakayama, H. and Y. Sawaragi (1984) Satisficing trade-off method for multiobjective programming, *Lecture Notes in Economics and Mathematical Systems* 229, Springer-Verlag, Heidelberg, 114-122.

Oral, M. and O. Kettani (1989) Modelling the process of multiattribute choice, *Journal of the Operational Research Society*, vol. 40, no 3, 281-291.

Paelinck, J. (1978) QUALIFLEX, a flexible multiple criteria method, *Economic Letters* 3, 193-197.

Pekelman, D. and S. Sen (1974) Mathematical programming models for the determination of attribute weights, *Management Science* 20, 1217-1229.

Philip, J. (1972) Algorithms for the vector maximization problem, *Mathematical Programming* 2, 207-229.

Pollak, R.A. (1967) Additive von Neumann-Morgenstern utility functions, *Econometrica* 35, 485-494.

Pomerol, J.-Ch. and S. Barba-Romero (1993) *Choix multicritere dans l'entreprise: Principe et pratique*, Hermes, Paris.

Pruitt, D. (1981) Negotiation behavior, *Academic Press*, New York.

Raiffa, H. (1969) Preferences for multi-attributed alternatives, *Memorandum RM-5868-DOT RC*, Rand Corporation, Santa Monica, California.

Roubens, M. (1981) Preference relations on actions and criteria in multiple criteria decision making, *European Journal of Operational Research* 10, 51-55.

Roy, B. (1968) Classement et choix en presence de points de vue multiples: La methode ELECTRE, *R.I.R.O.* 8, 57-75.

Roy, B. (1974) Criteres multiples et modelisation des preferences: L'apport des relations de surclassement, *Revue d'Economie Politique* 84, 1-44.

Roy, B. (1978) ELECTRE III: Un algorithme de classements fonde sur une representation floue des preferences en presence de criteres multiples, *Cahiers du CERO* 20, 3-24.

Roy, B. (1985) *Methodologie multicritere d'aide a la decision*, Economica, Paris.

Roy, B. (1994) La quarantieme reunion du groupe: Une reunion comme les autres?, *Newsletter of the European Working Group "Multicriteria Aid for Decisions"* 2, 5, 1-2.

Roy, B. and P. Bertier (1973) La methode ELECTRE II: Une application au media-planning, in: Ross, M. (ed.), *OR* 72, North-Holland, Amsterdam, 291-302.

Roy, B. and D. Bouyssou (1993) *Aide multicritere a la decision: Methodes et cas,* Economica, Paris.

Roy, B. and J.C. Hugonnard (1982) Ranking of suburban line extension projects on the Paris metro system by a multicriteria method, *Transportation Research* 16A, 301-312.

Roy, B. and J.M. Skalka (1984) ELECTRE IS: Aspects methodologiques et guide d'utilisation, *Document du LAMSADE* 30, Universite de Paris-Dauphine, Paris.

Saaty, T.L. (1980) *The analytic hierarchy process,* McGraw-Hill, New York.

Scharlig, A. (1985) *Decider sur plusieurs criteres,* Collection Diriger l'Entreprise, Presses Polytechniques Romandes.

Siskos, J. (1983) Analyse de systemes de decision multicritere en univers aleatoire, *Foundations of Control Engineering* 8, 193-212.

Siskos, J. and D. Yannacopoulos (1985) UTASTAR, an ordinal regression method for building additive value functions, *Investigacao Operacional* 5, 39-53.

Siskos, Y. and C. Zopounidis (eds) (1993) Multicriteria decision support systems, *Journal of Information Science and Technology* 2, 2, 107-108.

Siskos, J., G. Wascher and H.M. Winkels (1984) Outranking approaches versus MAUT in MCDM, *European Journal of Operational Research* 16, 270-271.

Siskos, Y., A. Spiridakos and D. Yannacopoulos (1993) MINORA: A multicriteria decision aiding system for discrete alternatives, *Journal of Information Science and Technology* 2, 2, in: Siskos, Y. and Zopounidis, C. (eds), *Multicriteria Decision Support Systems*, 136-149.

Spronk, J. (1981) *Interactive multiple goal programming,* Martinus Nijhoff, Boston.

Srinivisan, V. and A.D. Shocker (1973) Estimating the weights for multiple attributes in a composite criterion using pairwise judgments, *Psychometrika* 38, 473-493.

Steuer, R.E. and E.U. Choo (1983) An interactive weighted Tchebycheff procedure for multiple objective programming, *Mathematical Programming* 26, 326-344.

Steuer, R. E., J. Silverman and A.W. Whisman (1993) A combined Tchebycheff/Aspiration criterion vector interactive multiobjective programming procedure, *Management Science* 39, 10, 1255-1260.

Steuer, R.E., L.R. Gardiner, J. Gray (1994) A bibliographic survey of the activities and international nature of multiple criteria decision making, *Working Paper*, University of Georgia, Athens, Georgia.

Stewart, T.J. (1987) Pruning of decision alternatives in multiple criteria decision making, based on the UTA method for estimating utilities, *European Journal of Operational Research* 28, 79-88.

Stewart, T.J. (1992) A critical survey on the status of multiple criteria decision making: Theory and practice, *OMEGA International Journal of Management Science* 20, 5/6, 569-586.

Teghem, Jr., J., D. Dufrane, M. Thauvoye and P. Kunsch (1986) STRANGE: An interactive method for multiobjective linear programming under uncertainty, *European Journal of Operational Research* 26, 65-82.

Vincke, Ph. (1992) *Multicriteria decision-aid*, John Wiley & Sons Ltd., Baffins Lane.

von Neumann, J. and O. Morgenstern (1944) *Theory of games and economic behavior*, Princeton University Press, Princeton.

Wakker, P.P. (1989) *Additive representations of preferences,* Kluwer Academic Publishers, Dordrecht.

Wierzbicki, A.P. (1980) The use of reference objectives in multiobjective optimization, in: Fandel, G. and Gal, T. (eds), *Multiple Criteria Decision Making, Theory and Applications,* Springer-Verlag, Heidelberg, 468-486.

Winkels, H.M. (1980) Complete efficiency analysis for linear vector maximum systems: Theoretical background and an algorithm, *Document du LAMSADE* 13, Universite de Paris-Dauphine, Paris.

Young, F.W., J. de Leeuw and Y. Takane (1976) Regression with qualitative and quantitative variables: An alternating least squares method with optimal scaling features, *Psychometrika* 41, 505-529.

Yu, P.L. and M. Zeleny (1975) The set of all nondominated solutions in the linear cases and a multicriteria simplex method, *Journal of Mathematical Analysis and Applications* 49, 430-468.

Yu, W. (1992) ELECTRE TRI: Aspects methodologiques et manuel d'utilisation, *Document du LAMSADE* 74, Universite de Paris-Dauphine, Paris.

Zionts, S. (1992) The state of multiple criteria decision making: Past, present, and future, in: Goicoechea, A., Duckstein, L. and Zionts, S. (eds), *Multiple Criteria Decision Making,* Springer-Verlag, New York.

Zionts, S. and J. Wallenius (1976) An interactive programming method for solving the multiple criteria problem, *Management Science* 22, 652-663.

I. METHODOLOGY

A SHORT NOTE ON A METHODOLOGY FOR CHOOSING A DECISION-AID METHOD

Ph. VINCKE
Universite Libre de Bruxelles
Bld du Triomphe, C.P. 210/01
B-1050 Brussels-Belgium

Abstract

A methodology is proposed to be able to compare decision-aid methods on a sound basis. The methodology is illustrated on a small example, where a choice has to be made between two variants of ELECTRE II method.

1 Introduction

The purpose of this paper is to illustrate, on a small example, one of the points stressed in the "Manifesto of the new MCDA era" ([1]) and which is concerned with the need for a basic theory of MCDA procedures.
One of the important sentences of this Manifesto is:

"Although the great diversity of MCDA procedures may be seen as a strong point, it can also be a weakness. Up to now there is no possibility of deciding if one method makes more sense than another towards a specific problem situation. A systematic axiomatic analysis of decision procedures and algorithms is yet to be carried out".

We will consider in this paper two methods which are regularly used in applications and present a methodology to be able to compare them. The result contained in this paper is not new: it was presented, in a more formal and more complete manner, in "Exploitation of a crisp relation in a ranking problem" [3], but our purpose here is to establish the dialog with the other researchers and to try to convince them of the interest of this methodology. That is the reason why mathematical formulations have been avoided as much as possible, so that the reader can concentrate more on the approach than on formal notations.

2 Description of the methods

The methods we want to compare consist in exploiting a given relation S on a finite set A, in order to rank the elements of A from the best to the worse, with eventual tie: in the following, we will call "ranking" such a result (the mathematical term is complete preorder or, sometimes, weak order).

The procedures in question are in fact two variants of the second step of ELECTRE II method ([2]), where relation S is an outranking relation obtained from the aggregation of several criteria. In this step, alternatives which are on a same circuit of S are considered as equivalent and replaced by one representative. We denote by [a] the representative of all the alternatives which are on a same circuit as a, element of A.

Let A' be the set of these representatives (of course $[a] \equiv [b]$ iff a and b are on a same circuit of S). Given two distinct elements [a] and [b] of A', we define

$$[a]\ S'\ [b] \text{ iff } c\ S\ d,$$

for some c equivalent to a and some d equivalent to b, and we say in this case "[a] is better than [b]" or "[b] is worse than [a]".

Clearly, S' has no circuit; the two procedures for obtaining a ranking of the representatives are then the following.

PROC. 1.

- Put in the first class the elements of A' which are not worse than any other element.
- Remove the elements of the first class.
- Put in the second class the elements which are not worse than any other element in the remaining set.
- Remove the elements of the second class.
- Continue this procedure until all the elements are removed.

PROC. 2.

- Calculate the score of each element of A' (the score of [a] is the number of elements which are worse than [a] minus the number of elements which are better than [a]).
- Rank the elements in the decreasing order of their score.

In both procedures, the final ranking of the alternatives is the ranking of their representatives (remember that two alternatives which have the same representative are tie).

These two procedures are regularly described in the literature and used in applications. The problem is to know if it is possible to justifiy the choice of one of them instead of the other.

3 Methodology

The methodology we propose consists in defining a list of properties that should be respected and to verify which of them are satisfied by the procedures to compare. It is important to say here that the purpose is not to prove that one procedure is absolutely better than the other (in fact, the answer will depend on the list of properties).

The list of properties which is used in the comparison is not the description of a certain reality; it must be seen in a constructive perspective, as a tool for a deeper understanding of the procedures.

In a context of decision-aid, it is very important to have an idea of the implicit assumptions or properties of the methods which are used, in order to be able to confront them with the context of the problem and the attitude of the decision-maker.

We will now illustrate this methodology on our example.

4 A list of properties

Let us consider the following list of properties:

(i) if S has already the structure of a ranking, than the procedure should give this ranking;

(ii) if it is possible to obtain a ranking from S without deleting any information contained in S, then the procedure should propose such a ranking; in particular,

- (ii1) if S is transitive, the procedure should give a ranking containing S;
- (ii2) if the transitive closure of S is a ranking the procedure should give this ranking;
- (ii3) if P has no cycle, the procedure should give a ranking containing P, where P is defined by a P b iff a S b and not b S a:

(iii) the procedure should be independent of the labels of the alternatives;

(iv) if a new information is added, which is favourable to alternative a, than the position of a in the final ranking should not be depreciated; more precisely, if a new a relation S' is obtained from S by adding the fact that a S' b (everything else being unchanged). than the place of a in the ranking associated to S' should be at least as good as its place in the ranking associated to S.

(v) if the class of best elements in the ranking associated to S is removed. and if the procedure is applied to the remaining elements. then the ranking of those remaining elements should be unchanged.

5 Theorem (proof in [3])

a) Procedure 1 satisfies all the properties of the above list.
b) Procedure 2 satisfies properties (i). (ii2). (iii) but violates all the other properties of the above list.

6 Comments

The main interest of this kind of theorem is not the fact that. for the given list of properties. procedure 1 clearly dominates procedure 2. (It will often happen that there is no dominance between the methods which are compared). It is more in the information it gives on the specificities of each procedure.

With such an information. the choice of a method is not completely arbitrary: it can be justified on a sound basis.

7 Conclusions

A methodology has been proposed to be able to give arguments for the choice of a decision-aid method. It consists in establishing a list of properties which should be satisfied and in trying to find a method which has these properties. Three possibilities can occur:

a) either no method can simultaneously satisfy the given properties: this leads to an "impossibility theorem" which is often very instructive; it will imply a reflexion on the "rationality" of the properties, an analysis of variants of the properties in order to try to obtain a possibility theorem and, in any case, a better understanding of the situation (the most famous example is Arrow's theorem);

b) either several methods simultaneously satisfy the properties: the interest is then to classify them in function of additional properties in order to obtain a typology of the methods, which is fundamental in the construction of a theory for decision-aid;

c) or one and only one method simultaneously satisfies the properties: this leads to a "characterization theorem" which is of course interesting for a deeper understanding of the method.

References

[1] "*Manifesto of the new MCDA era*", in Journal of Multi-Criteria Decision Analysis, 2(3), pp. 125-128, 1993.

[2] Roy, B. and Bertier, P. "*La méthode ELECTRE II, une application au media-planning*", in O.R. 72, M. Ross (ed.), North-Holland, pp. 291-302, 1973.

[3] Vincke, Ph. "*Exploitation of a crisp relation in a ranking problem*", Theory and Decision, 32, pp. 221-240, 1992.

A NOTE ON THE 'MIN IN FAVOR' CHOICE PROCEDURE FOR FUZZY PREFERENCE RELATIONS

Denis BOUYSSOU
ESSEC

Abstract

This note deals with the problem of choosing on the basis of a fuzzy preference relation. In the area of MCDM, such a problem occurs with outranking methods using fuzzy relations such as ELECTRE III. In this note, we study a choice procedure based on the 'min in favor'. We show that it is the only one to satisfy a system of three independent axioms.

Keywords: Decision Making, Fuzzy Preferences, Choice, Outranking Methods, MCDM.

I - Introduction

Let A be a finite set of objects with at least two elements. We interpret the elements of A as alternatives among which a choice is to be made taking into account different points of view, *e.g.* several criteria or the opinion of several voters. A common practice in such a situation is to associate with each ordered pair (a, b) of alternatives a number indicating the strength or the credibility of the proposition "a is at least as good as b", *e.g.* the sum of the weights of the criteria favoring a or the percentage of voters declaring that a is preferred or indifferent to b. This leads to a fuzzy (large) preference relation on A. In the area of MCDM, ELECTRE III (Roy (1978)) is a typical illustration of such a process. It is well-known that, when the different points of view taken into account are conflictual, such a fuzzy preference relation does not, in general, possess "nice properties". In fact, Bouyssou (1994) has shown that any reflexive fuzzy binary relation can be obtained as the result of ELECTRE III. Choosing alternatives on the basis of such information is therefore far from being an easy task. This calls for the study of choice procedures for fuzzy preference relations.

A fuzzy (binary) relation on a set A is a function R associating with each ordered pair of alternatives $(a, b) \in A^2$ an element of [0, 1]. A fuzzy binary relation R on A is said to be reflexive if R(a, a) = 1 for all $a \in A$. The set of all fuzzy reflexive relations on A will be denoted by F(A). With ELECTRE III in mind, it will be supposed here any reflexive fuzzy binary relation on A may be the basis for choice. Therefore, we define a choice procedure for fuzzy preference relations (on a set A) as function associating a nonempty subset of A, the "choice set", with each fuzzy reflexive binary relation on A.

P.M. Pardalos et al. (eds.), Advances in Multicriteria Analysis, 9–16.

In this note, we study "choice procedures" instead of the more classical notion of "choice functions", *i.e.* functions associating a choice set with any subset of A. This is due to the fact that, following Sen (1993), we shall not make use of any "consistency property" of choice when the set of alternatives shrinks or expands (on such properties see, *e.g.*, Aizerman (1985) or Moulin (1985)).

If a fuzzy relation R ∈ F(A) is such that R(a, b) ∈ {0, 1}, for all a, b ∈ A, we say that R is crisp. In this case, we write a R b instead of R(a, b) = 1. We denote by U(A) the set of all crisp (Unfuzzy) relations in F(A), *i.e.* the set of all crisp reflexive binary relations on A. Throughout this note we adhere to the terminology of Roubens and Vincke (1985) concerning crisp relations.

The classical problem of defining "reasonable" choice procedures for crisp relations is not an easy one. It has generated numerous studies, in particular in the case of tournaments (*i.e.*, crisp, asymmetric and complete binary relations, see Moulin (1986)). This difficulty is largely due to the fact that when a crisp preference relation is not complete and/or has cycles in its asymmetric part, the very notion of a "good" alternative is not easy to define. Such relations are commonly found in Social Choice Theory (see McGarvey (1952) or Deb (1976)).

Turning now to fuzzy preference relations, the situation appears even more complex. This increased complexity stems in particular from:

- the possibility to generalize to the fuzzy case the classical properties of crisp binary relations (completeness, transitivity, absence of circuits, etc.) in many different ways, (see, *e.g.*, Barrett and Pattanaik (1989), Billot (1991), Dutta *et al.* (1986), Jain (1990), Montero and Tejada (1987), Ovchinnikov (1990) or Perny (1992));
- the difficulty to define in a consistent way the symmetric and asymmetric parts of a fuzzy relation and, hence, to separate strict preference from indifference (see, *e.g.*, Fodor (1991), Ovchinnikov and Roubens (1991, 1992) or Perny and Roy (1992));
- the difficulty to interpret the meaning of the numbers R(a, b): are they a cardinal measure of credibility or do they only represent an ordinal information on credibility ? (see, *e.g.*, Barrett *et al.* (1992), Basu *et al.* (1992) or Perny (1992)).

This note will not attempt to answer these difficult questions which are still widely open. Its purpose is to characterize what appears to be a "reasonable" choice procedure in the particular case in which the meaning of fuzziness is strictly "ordinal". This somewhat radical interpretation of fuzziness allows to considerably simplify the problem of defining "reasonable" choice procedures. This is because, within this interpretation, the answers to the first two questions are not so crucial.

II - Some properties of choice procedures

Consider a fuzzy relation R ∈ F(A), *i.e.* a function associating with each ordered pair of alternatives (a, b) ∈ A^2 an element of [0, 1]. Suppose that R(a, b) = 0.2 and R(c, d) = 0.8.

Should we conclude that the proposition "c is at least as good as d" is four times more credible than the proposition "a is at least as good as b" ? In some situations, *e.g.* when the numbers R(a, b) represent a proportion of voters or of criteria favoring the proposition "a is at least as good as b", this may seem reasonable and a choice procedure should take into account such considerations (for an example of such a procedure see Bouyssou (1992a)). In other situations, *e.g.* when the fuzzy relation has been obtained on a purely introspective basis or when the weights of the criteria only reflect an ordinal information about their respective importance, this may well lead to a somewhat unrealistic preference model. Many attempts have been made to propose an "ordinal" theory of fuzziness (see, *e.g.*, Goguen (1967) or, more recently, Basu *et al.* (1992)). In this note we shall remain in the ordinary framework of the theory of fuzzy sets but impose that a choice procedure should only make use of the underlying ordinal information on credibility conveyed by a fuzzy preference relation.

We say that a choice procedure C is **ordinal** if, for all $R \in F(A)$ and all strictly increasing and one-to-one transformation ϕ from [0, 1] to [0, 1],
$C(R) = C(\phi[R])$,
where $\phi[R]$ is the element of F(A) such that $\phi[R](c, d) = \phi(R(c, d))$ for all $c, d \in A$.

It is clear that an ordinal choice procedure does not make use of the cardinal properties of the numbers R(a, b). Many ordinal choice procedures can be envisaged. Let us mention one of them that has often been discussed in the literature and may be seen as a direct extension to the fuzzy case of the classical notion of the "greatest elements" of a crisp preference relation (see Switalski (1988)). Let $R \in F(A)$ and, for all $a \in A$, define, using the same notation as in Barrett *et al.* (1990), the 'min in Favor' score of alternative a letting:

$$mF(a, R) = \min_{c \in A \setminus \{a\}} R(a, c).$$

A clearly ordinal choice procedure is defined by:
$C_{mF}(R) = \{a \in A: mF(a, R) \geq mF(b, R) \text{ for all } b \in A\}$.
Many other ordinal choice procedures can be envisaged (see Barrett *et al.* (1990)).

A choice procedure for fuzzy relations should generate "reasonable choices" when applied to crisp relations. Given a crisp relation R on A define $G(R) = \{a \in A: a\,R\,b \text{ for all } b \in A\}$ as the subset of the R-greatest elements in A. Unless the crisp relation R has "nice properties", the set G(R) will often be empty. When it is not however, there is little interest in choosing alternatives outside G(R), since these alternatives are "at least as good" as every other alternatives in A. This is the *raison d'être* of the following axiom.

We say that a choice procedure C is **faithful** if, for all $R \in F(A)$,
$[R \in U(A) \text{ and } G(R) \neq \emptyset] \Rightarrow C(R) \subseteq G(R)$.

It is not difficult to see that C_{mF} is indeed faithful. A more restrictive version of faithfulness has been introduced by Barrett *et al.* (1990).

Faithfulness imposes a constraint on the result of a choice procedure when applied to (some) crisp relations. Ordinality imposes that the result of choice procedure is identical when applied to two "ordinally-equivalent" relations, *i.e.* to two relations R, S ∈ F(A) such that, for all a, b ∈ A, R(a, b) = ϕ(S(a, b)) for some strictly increasing and one-to-one transformation ϕ on [0, 1]. It should be noticed that no relation in F(A)\U(A) can be "ordinally-equivalent" to a relation in U(A) since only one-to-one transformations are invoked by ordinality. Thus, these two axioms impose very few constraints on the desirable behavior of a choice procedure when applied to fuzzy relations outside U(A). In particular, they leave room for "discontinuities", which seem rather paradoxical. Let us illustrate the possibility of discontinuities on a simple example involving a crisp relation and an "almost crisp" one. Consider the relations R and R′ on A = {a, b, c} defined by the following tables (to be read from row to column):

R	a	b	c
a	1	1	1
b	0	1	0
c	0	0	1

R′	a	b	c
a	1	1	λ
b	0	1	0
c	0	0	1

where $0 < \lambda < 1$.

It is easy to see that R is crisp and that G(R) = {a}. Let C be a faithful choice procedure. We have C(R) = {a}. Even if C is ordinal, it may happen that a ∉ C(R′) whatever the value of λ. As a result C(R)∩C(R′) will be empty even when R′ is arbitrarily "close" to R. Our final axiom is designed to prevent such situations.

Consider a sequence of valued relations (R_i ∈ F(A), i = 1, 2, ...). We say that this sequence converges to converges to R ∈ F(A) if, for all $\varepsilon \in \mathbb{R}$ with $\varepsilon > 0$, there is an integer k such that, for all $j \geq k$ and all a, b ∈ A, $|R_j(a, b) - R(a, b)| < \varepsilon$.

A choice procedure C is said to be **continuous** if, for all R ∈ F(A) and all sequences (R_i ∈ F(A), i = 1, 2, ...) converging to R,
[a ∈ C(R_i) for all R_i in the sequence] ⇒ [a ∈ C(R)].

Our definition of continuity implies that an alternative that is always chosen with fuzzy relations arbitrarily close to a given relation should remain chosen with this relation. It is not difficult to see that C_{mF} is continuous.

The result presented in the next section combines ordinality and continuity. This may appear awkward since ordinality implies that the cardinal properties of the numbers R(a, b) should not be used whereas continuity involves a measure of distance between fuzzy relations using these properties. In presence of ordinality, it would be more satisfactory to formulate a continuity requirement in purely ordinal terms, *e.g.* using a notion of distance between fuzzy relations based on the crisp relations in terms of credibility that they induce on AxA. Though it is fairly easy to reformulate our continuity axiom in this way, such a reformulation would involve a significant loss of intuitive appeal and would lead us outside the traditional framework of the theory of fuzzy sets. In this note, we do not pursue any further in this direction.

Ordinality and continuity impose severe restrictions on a choice procedure. These restrictions are easily understood using the notion of λ-cut. Let $R \in F(A)$ and $\lambda \in [0, 1]$. The λ-cut of R is the crisp relation $R_\lambda \in U(A)$ such that, for all a, b $\in$ A, a R_λ b if and only if $R(a, b) \geq \lambda$. It is not difficult to see that, for any $\lambda \in (0, 1]$, there is a sequence of one-to-one and strictly increasing transformations (ϕ_i, i = 1, 2, ...) on [0, 1] such that $\phi_i[R]$ converges to R_λ. Thus, ordinality and continuity implies that alternatives that are chosen with a fuzzy relation should also be chosen with all of its (strictly positive) λ-cuts. The result in the following section is based on this simple observation coupled with the fact that, since λ-cuts are crisp relations, the result of a choice procedure with such relations may be constrained by faithfulness.

III - Results and Discussion

The main purpose of this note is to prove the following:

Proposition
C_{mF} is the only choice procedure that is ordinal, continuous and faithful.

Proof
We already observed that the C_{mF} is ordinal, continuous and faithful.
Let us now show that if a choice procedure C is ordinal, continuous and faithful then, for all $R \in F(A)$ and all a, b $\in$ A:

$mF(a, R) > mF(b, R) \Rightarrow b \notin C(R)$ and (α)

$mF(a, R) = mF(b, R)$ and $b \in C(R) \Rightarrow a \in C(R)$. (β)

which will complete the proof.
In contradiction with (α), suppose that $mF(a, R) > mF(b, R)$ and $b \in C(R)$ for some ordinal, continuous and faithful choice procedure C, some $R \in F(A)$ and some a, b $\in$ A. Let $\lambda \in (mF(b, R), mF(a, R))$ and consider any sequence of strictly increasing and one-to-one transformations (ϕ_i, i = 1, 2, ...) on [0, 1] such that:

$$\lim_{i \to \infty} \phi_i(x) = \begin{cases} 1 \text{ if } x \geq \lambda \text{ and} \\ 0 \text{ otherwise.} \end{cases}$$

(A simple example of such a sequence of functions is:

$$\phi_i(x) = \begin{cases} \dfrac{\lambda^{1/i} + (x-\lambda)^{1/i}}{\lambda^{1/i} + (1-\lambda)^{1/i}} & \text{if } x \geq \lambda \text{ and} \\ \dfrac{\lambda^{1/i} - (\lambda - x)^{1/i}}{\lambda^{1/i} + (1-\lambda)^{1/i}} & \text{otherwise.)} \end{cases}$$

By construction, the sequence ($\phi_i[R]$, i = 1, 2, ...) converges to the λ-cut R_λ of R. It is clear that $a \in G(R_\lambda)$ and $b \notin G(R_\lambda)$. By ordinality and continuity we know that $b \in C(R_\lambda)$ which contradicts faithfulness and proves (α).
In order to prove (β), suppose that $mF(a, R) = mF(b, R)$ and $b \in C(R)$, for some ordinal, continuous and faithful choice procedure C, some $R \in F(A)$ and some a, b $\in$ A.

Since $b \in C(R)$, we know, using part (α), that $mF(b, R) \geq mF(c, R)$ for all $c \in A$. Consider the sequence ($R_i \in F(A)$, $i = 1, 2 \ldots$) converging to R, where R_i is identical to R except that $R_i(b, c) = Max(0 ; R(b, c) - 1/i)$ for all $c \in A \setminus \{b\}$ and $R_i(a, d) = Min(1 ; R(a, d) + 1/i)$ for all $d \in A \setminus \{a\}$.
By construction, we have $mF(a, R_i) > mF(c, R_i)$ for all $c \in A \setminus \{a\}$. Since $C(R_i)$ is nonempty, (α) implies that $C(R_i) = \{a\}$ for all R_i in the sequence. Thus continuity implies $a \in C(R)$. This proves (β) and completes the proof. □

We conclude this note with some remarks.

a) The three axioms that we used to characterize C_{mF} are independent as shown by the following examples (we use the notations of Barrett *et al.* (1990).

i- Let C_{SF} be the ("Sum in Favor") choice procedure defined as:
$C_{SF}(R) = \{a \in A: SF(a, R) \geq SF(b, R) \text{ for all } b \in A\}$ where

$$SF(a, R) = \sum_{c \in A \setminus \{a\}} R(a, c).$$

This choice procedure is continuous and faithful but not ordinal.

ii- Let C_{MA} be the ("Max Against") choice procedure defined as:
$C_{MA}(R) = \{a \in A: MA(a, R) \leq MA(b, R) \text{ for all } b \in A\}$ where

$$MA(a, R) = \max_{c \in A \setminus \{a\}} R(c, a).$$

This choice procedure is ordinal and continuous but not faithful.

iii- Let C_L be the choice procedure defined by:
$C_L(R) = \{a \in C_{mF}(R): MF(a, R) \geq MF(b, R) \text{ for all } b \in C_{mF}(R)\}$, where

$$MF(a, R) = \max_{c \in A \setminus \{a\}} R(a, c).$$

This choice procedure is ordinal and faithful. It is not difficult to see that it is not continuous.

b) The mF score can be used not only to define a choice procedure – selecting the alternatives with the highest score – but also to rank order alternatives according to their scores. Independent characterizations of such a ranking procedure have been obtained by Pirlot (1992, 1994) and Bouyssou (1992b). Though the ranking procedure for fuzzy preference relations based on the mF score has many interesting properties, applying this ranking procedure to a complete and transitive crisp relation leads a complete and transitive crisp relation that may be different from the first one, thus violating a possible interpretation of "faithfulness" for ranking procedures (it should be noted that a "faithful" ranking procedure based on the mF score can be obtained by an iterative use of C_{mF}; such a "faithful" ranking procedure would not behave very well in terms of monotonicity however – see Perny (1992)). Thus the mF score seems to be more adapted to choice problems than to ranking problems. Indeed, we have shown that if the reasonableness of ordinality, faithfulness and continuity is admitted then C_{mF} is the only "reasonable" choice procedure.

c) Let us finally mention that a truly satisfactory fuzzy model of preferences probably lies in between strict ordinality – as was supposed here – and full cardinality allowing all possible operations on the numbers R(a, b). A precise formalization of such intermediate situations certainly deserves further investigations.

References

Aizerman, M.A. (1985), New problems in the general choice theory: review of research trend, *Social Choice and Welfare*, **2**, 235-282.

Barrett, C.R. and Pattanaik, P.K. (1989), Fuzzy sets, preference and choice: some conceptual issues, *Bulletin of Economic Research*, **41**, 229-253.

Barrett, C.R., Pattanaik, P.K. and Salles, M. (1990), On choosing rationally when preferences are fuzzy, *Fuzzy Sets and Systems*, **34**, 197-212.

Barrett, C.R., Pattanaik, P.K. and Salles M. (1992), Rationality and aggregation in an ordinally fuzzy framework, *Fuzzy Sets and Systems*, **49**, 9-14.

Basu, K., Deb, R. and Pattanaik, P.K. (1992), Soft sets: an ordinal formulation of vagueness with some applications to the theory of choice, *Fuzzy Sets and Systems*, **45**, 45-58.

Billot, A. (1991), Aggregation of preferences: the fuzzy case, *Theory and Decision*, **30**, 51-93.

Bouyssou, D. (1992a), A note on the sum of differences choice function for fuzzy preference relations, *Fuzzy Sets and Systems*, **47**, 197-202.

Bouyssou, D. (1992b), A Note of the 'Min In Favor' Ranking Method for Valued Preference Relations, in M. Cerny, D. Glükaufová and D. Loula (Eds), *Multicriteria Decision Making. Methods - Algorithms - Applications*, Czechoslovak Academy of Sciences, Prague, 16-25.

Bouyssou, D. (1994), *Outranking relations: do they have special properties ?*, Submitted to the Journal of Multiple criteria Decision Analysis.

Deb, R. (1976), On constructing generalized voting paradoxes, *Review of Economic Studies*, **43**, 347-351.

Dutta, B., Panda, S. and Pattanaik, P.K. (1986), Exact choice and fuzzy preferences, *Mathematical Social Sciences*, **11**, 53-68.

Fodor, J. (1991), Strict preference relations based on weak t-norms, *Fuzzy Sets and Systems*, **43**, 327-336.

Goguen, J.A. (1967), L-fuzzy sets, *Journal of Mathematical Analysis and Applications*, **18**, 145-174.

Jain, N. (1990), Transitivity of fuzzy relations and rational choice, *Annals of Operations Research*, **23**, 265-278.

McGarvey, D.C. (1953), A theorem on the construction of voting paradoxes, *Econometrica*, **21**, 608-610.

Montero, F.J. and Tejada, J. (1987), Fuzzy preferences in decision making, in B. Bouchon and R.R. Yager (eds.), *Uncertainty in Knowledge-Based Systems*, Springer-Verlag, Berlin, 144-150.

Moulin, H. (1985), Choice functions over a finite set: a summary, *Social Choice and Welfare* **2**, 147-160.

Moulin, H. (1986), Choosing from a tournament, *Social Choice and Welfare*, **3**, 271-291.

Ovchinnikov, S.V. (1990), Modelling valued preference relations, in J. Kacprzyck and M. Fedrizzi (eds.), *Multiperson Decision Making Models using Fuzzy Sets and Possibility Theory*, Kluwer Academic Publishers, 143-154.

Ovchinnikov, S.V. and Roubens, M. (1991), On strict preference relations, *Fuzzy Sets and Systems*, **43**, 319-326.

Ovchinnikov, S.V. and Roubens, M. (1992), On fuzzy strict preference, indifference, and incomparability relations, *Fuzzy Sets and Systems*, **49**, 15-20.

Perny, P. (1992), *Modélisation, agrégation et exploitation de préférences floues dans une problématique de rangement*, Unpublished Doctoral Dissertation, Université de Paris-Dauphine, Paris.

Perny, P. and Roy, B. (1992), The use of fuzzy outranking relations in preference modelling, *Fuzzy Sets and Systems*, **49**, 33-53.

Pirlot M. (1992), Ranking procedures for valued preference relations: a characterization of the 'min' procedure, *Cahiers du CERO*, **34**, 233-241.

Pirlot, M. (1994), *A characterization of "min" as a procedure for exploiting valued preference relations and related results*, forthcoming in the Journal of Multiple Criteria Decision Analysis.

Roubens, M. and Vincke, Ph. (1985), *Preference Modelling*, Springer Verlag, Berlin.

Roy, B. (1978), ELECTRE III : un algorithme de classement fondé sur une représentation floue des préférences en présence de critères multiples, *Cahiers du CERO*, **20**, 3-24.

Sen, A.K. (1993), Internal consistency of Choice, *Econometrica*, **61**, 495-521.

Switalski, Z. (1988), Choice functions associated with fuzzy preference relations, in J. Kacprzyk and M. Roubens (eds.), *Non-conventional Preference Relations in Decision Making*, Springer-Verlag, Berlin.

Acknowledgments. I wish to thank Patrice Perny and Marc Pirlot for helpful discussions on the subject of this note. I am indebted to an anonymous referee for extremely valuable suggestions.

Denis BOUYSSOU
ESSEC
BP 105
F-95021 Cergy - France

e-mail: bouyssou@ped.essec.fr

ELICITING INFORMATION CONCERNING THE RELATIVE IMPORTANCE OF CRITERIA

Vincent Mousseau
LAMSADE - Université Paris-Dauphine
Place du maréchal De Lattre de Tassigny
75 775 Paris cedex 16 - FRANCE
Email: mousseau@lamsade.dauphine.fr

Abstract: The notion of Relative Importance of Criteria (RIC) is central in the domain of Multiple Criteria Decision Aid (MCDA). It aims at differentiating the role of each criterion in the construction of comprehensive preferences, thus allowing to discriminate among pareto-optimal alternatives. In most aggregation procedures, this notion takes the form of importance parameters.

The acquisition of information concerning the RIC may be supported by Elicitation Techniques for Importance Parameters (ETIP). The design of such techniques should account for both the meaning that each aggregation confers on its parameters and the decision makers' (DMs) understanding of the notion of RIC. More precisely, ETIPs should be able to provide a good fit between the way the analyst uses the DM's assertions in the model and the information that he/she expresses through his/her statements.

In this paper, we present an ETIP adapted to the ELECTRE methods that proceeds by means of pairwise comparisons of fictitious alternatives. Implemented in a software program called DIVAPIME, this ETIP supports the elicitation of variation intervals for the ELECTRE methods' preferential parameters.

Keywords: MCDA, Importance of Criteria, Weights, Elicitation, ELECTRE Methods.

Introduction

When the analysis of a decision problem is grounded in the definition of a set of criteria, it is difficult to discriminate between alternatives whose evaluations on several criteria are in conflict. Multiple criteria preference modeling requires that the analyst obtains from the decision maker (DM) some preference information so as to discriminate between pareto-optimal alternatives.

P.M. Pardalos et al. (eds.), Advances in Multicriteria Analysis, 17–43.

A classical approach to Multiple Criteria Decision Aid (MCDA) consists of linking restricted preferences (corresponding to the n criteria) with the comprehensive preferences (taking all criteria into account) through a so called Multiple Criteria Aggregation Procedure (MCAP). In the MCAP, all criteria are not supposed to play the same role; the criteria are commonly said not to have the same importance. This is why there are parameters in the MCAP that aim at specifying the role of each criterion in the aggregation of evaluations. We will call such parameters importance parameters. They aim at introducing preferential information concerning the importance that the DM attaches to the points of view modelled by the criteria.

The nature of these parameters varies across MCAPs. The way of formalizing the relative importance of each criterion differs from one aggregation model to another. All this is done, for instance, by means of:

- scaling constants in Multiattribute Utility Theory (see [Keeney & Raiffa 76,93]),
- a weak-order on F in lexicographic techniques or a complete pre-order on F in the ORESTE method (see [Roubens 82]),
- intrinsic weights in PROMETHEE methods (see [Brans et al. 84]),
- intrinsic weights combined with veto thresholds in ELECTRE methods (see [Roy 91]),
- eigen vectors of a pairwise comparison matrix in AHP method (see [Saaty 80]).

Many authors have studied the problem of the elicitation of the relative importance of criteria (a critical overview may be found in [Mousseau 92]), but few of them have tried to give a precise definition of this notion. A careful analysis of this notion is still needed to build theoretically valid elicitation techniques. This paper aims at highlighting some of the difficulties that may be encountered when eliciting information concerning the relative importance of criteria and presents a way of eliciting such information when preferences are modeled through an outranking relation based on a concordance principle (see [Roy 91]).

In the first section, we will state precisely what the information underlying the notion of importance of criteria is. The second section will be devoted to the analysis of the meaning of an elicitation process. This will lead us to specify some basic requirements for importance parameters elicitation techniques. In the last section, we will present a technique, called DIVAPIME, in order to define a polyhedron of acceptable values for importance parameters in an ELECTRE type method.

1. What does the notion of the relative importance of criteria cover ?

What does a decision-maker mean by assertions such as "criterion g_j is more important than criterion g_i," "criterion g_j has a much greater importance than criterion g_i?", etc. Let us recall, by way of comparison, that the assertion "b is preferred to a" reflects the fact that, if the decision-maker must choose between the alternatives b and a, he is supposed to decide in favor of b. So, it is possible to test whether this assertion is valid or not. There is no similar possibility for comparing the Relative Importance of Criteria (RIC). Moreover, the way this notion is taken into account within the framework of the different models mentioned above by means of importance parameters reveals significant differences in what this notion deals with.

The first statement we should make when trying to analyse the notion of RIC is the following: the information underlying this notion is much richer than that contained in the importance parameters of the various multicriteria models. In fact, these parameters are mainly scalars and constitute a simplistic way of taking RIC into account, as this notion is by nature of a functional type.

In the comparison of two alternatives a and b, when one or several criteria are in favor of a and one or several others in favor of b, the way each MCAP solves this conflict and determines a comprehensive preference (i.e., taking all criteria into account) denotes the importance attached to each criterion (and to the logic of the aggregation used). Thus, the result of such conflicts (i.e., the comprehensive preference situation between a and b) constitutes the elementary data providing information on the relative importance of the criteria in conflict.

When we analyse the RIC notion, it appears that this notion represents a certain form of regularity in the link between restricted and overall preferences. In order to delimit exhaustively the importance of a criterion, we should analyse the contribution of any preference at the restricted level of a criterion to the comprehensive level (for each pair of alternatives). Nevertheless, when two alternatives are indifferent on criterion g_j, the comprehensive preference situation will generally give no significant information concerning the importance of this criterion.

Let us introduce some basic notations:

F={1,2, ..., n} a familly of n criteria g_1, g_2, ..., g_n build so as to evaluate the alternatives contained in a given set (denoted A). Considering the imprecision of the evaluations of alternatives on criteria, it is usually considered that g_1, g_2, ..., g_n are pseudo-criteria (see [Vincke 90]), i.e., **such that:**

$$\begin{cases} aP_jb \Leftrightarrow g_j(a) > g_j(b) + p_j \\ aQ_jb \Leftrightarrow q_j < g_j(a) - g_j(b) \leq p_j \\ aI_jb \Leftrightarrow |g_j(a) - g_j(b)| \leq q_j \end{cases}$$

where - q_j, the indifference threshold, represents the maximum difference of evaluation compatible with an indifference situation;

- p_j, the preference threhold, represents the minimum difference of evaluation compatible with a preference situation;
- aP_jb is to be interpreted as a preference for a over b on criterion g_j;
- aI_jb is to be interpreted as indifference between a and b on criterion g_j;
- aQ_jb is to be interpreted as an hesitation between the two preceding situations.

We will call P_j the partial preference relation on the jth criterion (the same terminology holds for I_j and Q_j). (I_j, Q_j, P_j) defines a pseudo-order.

The comprehensive preferences are modelled through three preference relations: P a preference relation (asymmetric, irreflexive), I an indifference relation (symmetric, reflexive), R an incomparability relation (symmetric, reflexive). For every pair of alternatives (a, b), one and only one of the four following assertions is valid:

$$\left[\begin{array}{l} aPb \\ aIb \\ bPa \\ aRb \end{array}\right. \qquad [1]$$

Let us consider the outranking relation S=P∪I, aSb being interpreted as "*a is at least as good as b*". As aPb ⇒ aSb ⇒ not[bSa], the four assertions of system [1] correspond to the following assertions:

$$\left[\begin{array}{l} aPb \\ aSb \text{ and not}[aPb] \\ bPa \\ \text{not}[aSb] \text{ and not}[bPa] \end{array}\right. \qquad [2]$$

When the preference model is of a (I, P, R) type, it does not seem restrictive to delimit the information concerning the relative importance of a criterion using only preference and outranking relations (see system [2]). On such a basis, we will assume that the empirical content of the RIC notion refers to the nature and variety of cases in which a partial preference on a given criterion g_j leads us (i) to accept the same preference on the comprehensive level, or (ii) to accept only an outranking in the same direction, or (iii) to refuse the inverse preference. In other words, the relative importance of criterion g_j is characterized by:

- the set of "*situations*" in which aP_jb and aPb hold simultaneously,
- the set of "*situations*" in which aP_jb and aSb hold simultaneously,
- the set of "*situations*" in which aP_jb and not[bPa] hold simultaneously.

In this characterization, the contribution of criterion g_j to the preferences at the comprehensive level is considered only through the situations of strict preference at the restricted level of criterion g_j. This does not seem restrictive when all criteria are quasi-criteria (such that $p_j=q_j$), but it can become restrictive when F contains pseudo-criteria.

On the basis of the preceding considerations [Roy & Mousseau 95] propose a formal definition of the notion of RIC and a theoretical framework to analyse it. Within this framework, it clearly appears that the importance of criteria is taken into account in very different ways in the various aggregation procedures. In particular, this means that the values attached to importance parameters are meaningless as long as the aggregation rule in which they are used is not specified. Interesting theoretical proposals can also be found in [Podinovskii 88, 94].

2. What do we aim at when evaluating the importance of criteria ?

The elicitation of information about RIC is a crucial phase in a decision aid process. Basic assumptions concerning the nature of what is being done during this phase have an impact on the way to proceed to elicit such information.

2.1. The descriptivist and constructivist approaches to MCDA

The way we seek to give meaning to the notion of importance will differ according to which of these two approaches we adopt.

The *descriptivist approach* assumes that the way in which any two alternatives are compared on the comprehensive level (that is, taking all criteria into account) is well-defined in the decision-maker's mind before the modelling process begins. Moreover, it supposes that the modelling process does not modify such comparisons. The preference model chosen is intended to give an account of such pre-existing preferences as objectively as possible. Under these conditions, the role which devolves to each criterion as a function of its importance, is apprehended by the set of values attributed to the importance parameters. Thus, it is the capacity of a model to adjust to a well-defined, real-world situation which confers meaning on the notion of importance and allows to assign a numerical value to these parameters. Several authors would even talk in terms of estimating the numerical value of certain parameters, such as weight w_j. Such language is meaningless unless a true numerical value for w_j exists, in which case the goal would be to estimate this true value as precisely as possible. In such an approach, observed lability in elicited preferences (see [Fishhoff et al. 89] and [Weber & Borcherding 93]) are explained by the existence of several biases in the elicitation techniques.

[Beattie & Baron 91] argues that there is a "*distinction between true and estimated weights and it is possible that subjects' true weights remain constant at all times, but become distorted in the elicitation process.*"

The *constructivist approach*, on the contrary, assumes that preferences are not entirely pre-formed in the decision-maker's mind and that the very nature of the work involved in the modelling process (and, *a fortiori*, in decision-aid) is to specify and even to modify pre-existing elements. The multiple criteria aggregation procedure (MCAP), which underlies the preference model chosen, is thus nothing other than a set of rules deemed appropriate for aggregating the evaluations $g_j(a)$ and for building comprehensive preferences. Under these conditions, the numerical values assigned to importance parameters reflect a working hypothesis accepted for decision-aid. These are convenient numerical values with which it seems reasonable and instructive to work. The importance parameters, therefore, can be seen as keys which allow us to differentiate the role played by each criterion in the preference model selected for use. "True numerical values," to which we can refer to give meaning to the language of estimation, do not necessarily exist. Importance parameters and the numerical values we assign to them are, nonetheless, instruments for reasoning, investigating and communicating among the stakeholders in a decision-making process. The values (or interval of variation) for these parameters reflect, in the MCAP selected, a certain number of assertions expressed by the DM during the elicitation process (see [Roy 93], [Mousseau 93] and [Paynes et al. 92]).

2.2. Importance parameters elicitation methods

Developing Elicitation Techniques for Importance Parameters (ETIP) is an area of research that lies between the theoretical analysis of the notion of RIC and the empirical investigation of decision behavior (see figure 1). In fact, any ETIP should account both for the precise meaning of the importance parameters in the MCAP used and for the DMs behavior related in empirical studies. For this interaction to be pertinent, it is crucial that the way the analyst uses the DM's answers in the model should conform to the information that the DM expressed through his answers.

Schematically, two different components of an ETIP are generally distinguished: The *questioning procedure* specifies how information is collected from the DM, i.e., the questioning mode and the sequence of these questions. The *deduction method* uses information obtained by the questioning procedure, verifying if this information is compatible with the chosen MCAP so as to infer values for importance parameters.

The various ETIPs proposed in the literature (see [Mousseau 92] for a review) may be classified into two categories. In the first category, *direct ETIPs* require an

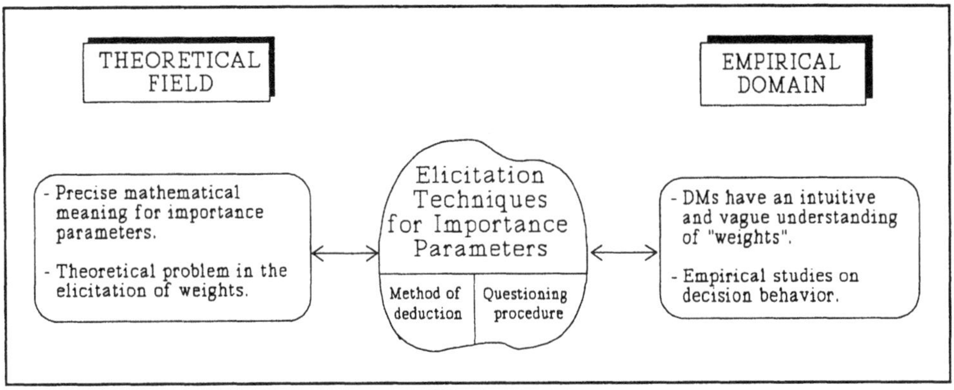

Figure 1

information on the concept of importance from the DM (direct evaluation of parameters, comparison of criteria in term of importance, etc.). The way the values for importance parameters are derived is defined independently of the aggregation rule in which these values will be used. Proceeding in this way, these ETIPs are not able to ensure that the information expressed in the DM's answers matches the use of this information in the MCAP.

On the contrary, *indirect ETIPs* explicitly intergrate the MCAP selected for use. The interaction with the DM is not based directly on the concept of importance but on indirect information (binary comparison or ranking of alternatives, for example) from which information concerning the RIC is inferred through the aggregation rule.

It follows from the preceeding considerations that any ETIP should satisfy two types of requirements:

i) *Any ETIP should explicitly refer to the MCAP that is used to model the DM's preferences*: The logic of the various MCAP implies different meanings for their importance parameters. This means that the result of conflicts between criteria is determined both by the values assigned to importance parameters and by the logic of the MCAP. The knowledge of the value of these parameters is not sufficient to discriminate between pareto optimal alternatives; thus the meaning of importance parameters is only defined in relation to the MCAP in which they are used (see section 1).

ii) *The way an ETIP interacts with the DM should account for his perception of the notion of RIC and for its limitations in perceiving and processing information*: Any ETIP proceeds through a phase of interaction with the DM. Many studies in behavioral science and experimental psychology are useful for defining questioning procedures, pertinent from the DMs' point of view (for an overview, see [Von Winterfeld & Edwards 86], [Paynes et al. 93]).

3. DIVAPIME: a way to elicit the importance of criteria in an MCAP based on a concordance principle

In this section, we present an indirect ETIP, implemented in software called DIVAPIME (**D**étermination d'**I**ntervalles de **VA**riation pour les **P**aramètres d'**I**mportance des **M**éthodes **E**lectre). This software supports the elicitation of all preferential parameters of the Electre methods (see [Roy 91]). However, the part concerning the elicitation of the importance coefficients may also apply to other MCAPs that build one or several outranking relations on a concordance concept such as PROMETHEE (see [Brans et al. 84]) or TACTIC (see [Vansnick 86]). DIVAPIME stems from the ETIP used in [Roy et al. 86] and presented in [Roy & Bouyssou 93]). Besides the implementation, our work consisted of restructuring and extending the questioning procedure to make it more precise, as well as improving the algorithmic aspects of the method.

3.1. The preferential parameters of the ELECTRE methods

ELECTRE methods build an outranking relation S, i.e., validating or invalidating, for any pair of alternatives (a,b), an assertion aSb, whose meaning is "*a is at least as good as b*". In the ELECTRE methods, preferences restricted to the significance axis of each criterion are defined through pseudo-criteria (see section 1). The indifference and preference thresholds (q_j and p_j) model the intra-criterion preferential information. They account for the imprecise nature of the evaluations $g_j(a)$. The q_j threshold specifies the largest difference of evaluation $g_j(a)$-$g_j(b)$ that preserves indifference between a and b (aI_jb); p_j represents the smallest difference $g_j(a)$-$g_j(b)$ compatible with a preference situation in favor of a (aP_jb). The weak preference relation Q_j should be interpreted as hesitating between opting for a preference or indifference situation.

At the comprehensive level of preferences, in order for the assertion aSb to be valid, two conditions should be verified:

- concordance: for an outranking aSb to be accepted, a "sufficient" majority of criteria should be in favor of this assertion,
- non-discordance: when the concordance condition holds, none of the criteria in the minority should oppose to the assertion aSb in "a too strong way".

Two types of importance parameters intervene in the construction of S:

- the set of importance coefficients (k_1,k_2, ..., k_n) takes into account the relative importance of coalitions of criteria and intervenes in the construction of a concordance index c(aSb) which is defined by:

$$c(aSb) = \frac{\sum_{i=1}^{n} k_k . c_i(aSb)}{\sum_{i=1}^{n} k_i} \quad \text{with} \quad c_j(aSb) = \begin{cases} = 1 & \text{if } aP_jb \text{ or } aQ_jb \text{ or } aI_jb \\ = 0 & \text{if } bP_ja \\ \in [0,1] & \text{if } bQ_ja \end{cases}$$

- the set of veto thresholds $(v_1(g_1), v_2(g_2), \ldots, v_n(g_n))$ is used in the discordance concept. $v_j(g_j)$ represents the greatest difference of evaluation $g_j(b)-g_j(a)$ compatible with the assertion aSb.

3.2. Determination of a polyhedron of admissible values for importance parameters

In this section, we present a technique which aims at determining a non-empty polyhedron of admissible values for $k=(k_1,k_2,\ldots\ldots,k_n)$ starting from linear inequalities on these coefficients. These inequalities come from DM's answers to binary comparisons of fictitious alternatives (see 3.3). We do not aim at determining a single vector of values k, but a set of vectors consistent with assertions expressed by the DM.

3.2.1. Formulating information through a segmented description

Let us suppose that criteria can be ordered by importance and let us renumber them so that $k_1<k_2<\ldots\ldots<k_n$. We want to specify an interval $]m_i,M_i[$ containing each k_i ($\forall i\neq 1$). Each of these intervals constitutes a segment in which the value of k_i may vary. We aim at building intervals $]m_i,M_i[$ such that m_i and M_i depend only on the k_j for $j<i$; Hence we will denote them $m_i(k_1,k_2,\ldots,k_{i-1})$ and $M_i(k_1,k_2,\ldots,k_{i-1})$. The system of inequalities is then:

$$\begin{aligned} m_2(k_1) &< k_2 < M_2(k_1) \\ m_3(k_1,k_2) &< k_3 < M_3(k_1,k_2) \\ m_4(k_1,k_2,k_3) &< k_4 < M_4(k_1,k_2,k_3) \\ &\vdots \\ m_{n-1}(k_1,k_2,\ldots\ldots,k_{n-2}) &< k_{n-1} < M_{n-1}(k_1,k_2,\ldots\ldots,k_{n-2}) \\ m_n(k_1,k_2,\ldots\ldots,k_{n-1}) &< k_n < M_n(k_1,k_2,\ldots\ldots,k_{n-1}) \end{aligned}$$

In what follows, $m_i(k_1,k_2,\ldots\ldots,k_{i-1})$ will be defined by the maximum of a list of lower bounds of k_i, and $M_i(k_1,k_2,\ldots\ldots,k_{i-1})$ by the minimum of a list of upper bounds of k_i.

Each of these lower or upper bounds takes the form of linear combination of k_j for $j<i$[1]. For example, the value of $m_5(k_1,k_2,k_3,k_4)$ can be $\{k_1+k_4,k_2+k_3\}$. So as to add a new inequality to the system[2], we proceed as follows: such an inequality may always be written as $\sum_{i=1}^{n}\alpha_i.k_i>0$, with $\alpha_i\in\mathbb{R}$. If imax is defined as the index of the greatest non-null coefficient α_i ($imax=Max\{i/\alpha_j=0\ \forall j>i\}$) ,we may rewrite the preceding inequality as $\alpha_{imax}.k_{imax}<\sum_{\substack{i\in F\\ i\neq imax}}-\alpha_i.k_i$. According to the sign of α_{imax}, this inequality specifies a new upper or lower bound for k_{imax}. It is then necessary to check that each lower bound of k_{imax} is still lower than each upper bound of k_{imax}. This verification may lead to adding new bounds. This step is called the saturation of the segmented description.

3.2.2. Saturation of the segmented description

At a given stage, adding an inequality may produce supplementary inequalities. If the added inequality specifies a new upper or lower bound for k_i so that $m_i(k_1,k_2,.....,k_{i-1})<k_i<M_i(k_1,k_2,.....,k_{i-1})$, by transitivity, it should hold:

$$m_i(k_1,k_2,.....,k_{i-1})<M_i(k_1,k_2,.....,k_{i-1}) \qquad [i]$$

If [i] is verified $\forall\ k_1,k_2,.....,k_{i-1}$, then no additional inequality should be generated; in the opposite case, this inequality [i] should be integrated into the system, and will produce one (or several) upper and/or lower bound(s) from which one (or several) additional inequalities might be generated.

Hence, saturating a segmented description consists of producing all possible (upper or lower) bounds from inequalities similar to [i] that are not verified $\forall\ k_1,k_2,.....,k_{i-1}$. The generation of a new bound can itself produce an additional bound. The saturation algorithm stops when no bound can be generated.

In certain cases, the addition of an inequality to the system can lead to an empty polyhedron. Such a situation occurs when a lower bound of a k_j is greater than one of the upper bounds of this k_j, $\forall\ k_1,k_2,.....,k_{j-1}$. This situation characterizes a contradiction between the last inequality integrated into the system and one or several others.

---

[1] The representation $m_i(k_1,k_2,...,k_{i-1})<k_i<M_i(k_1,k_2,...,ki-1)$ of the domain of variation of $k=(k_1,k_2,...,k_n)$ exists whether this domain is a polyhedron or not; the fonctions $m_i(k_1,k_2,.....,k_{i-1})$ and $M_i(k_1,k_2,.....,k_{i-1})$ may not be linear (see [Roy 70], chap. 10).

[2] This inequality comes from the answer to a pairwise comparison of alternatives (see 3.3.1).

The algorithm can detect such inconsistencies and determine which inequalities generate the contradiction[3]. In order for the polyhedron of admissible solutions not to become empty, it is necessary to delete one or several inequalities generating the inconsistency together with all bounds that have been generated by the bound(s) to be deleted.

The convergence of the saturation mechanism has been proved (see [Roy 70]) ; we are then sure to reach a saturated segmented description. Let us give, for explanatory purposes, some elements that justify the convergence of the algorithm: at the saturation step, the production of an additional bound comes from the fact that the inequality $m_i(k_1,k_2,.....,k_{i-1})<M_i(k_1,k_2,.....,k_{i-1})$ is not verified, $\forall k_1,k_2,.....,k_{i-1}$. In this case, the new bound will always be added on a segment k_j such that $k_j<k_i$ (indeed, $m_i(k_1,k_2,.....,k_{i-1})$ and $M_i(k_1,k_2,.....,k_{i-1})$ are functions of k_j such that $j<i$). It follows that such deductions can be carried out only a finite number of times and that the saturation should be computed linearly, starting from the segment corresponding to the most important criterion without any backtracking to saturated segments.

3.3. Questioning procedure

3.3.1. Questioning mode: pairwise comparisons of fictitious alternatives

As this ETIP is an indirect ETIP, the questioning mode does not directly refer to the concept of importance. We can, from the answers of the DM to the questions, infer information through the MCAP used. The questioning mode selected is a pairwise comparison of fictitious alternatives. For each question, the DM has to define the comprehensive preference situation between two evaluation vectors.

The fictitious alternatives involved in the comparisons are chosen so as to provide specific information. Moreover, they should be able to correspond to real alternatives (their evaluations should be plausible and respect possible statistical links between criteria). The questioning procedure is founded upon the following fictitious alternatives:

b_0: A reference alternative whose evaluations on each criterion are "average".

b_i: Alternatives whose evaluations are identical to b_0 on all criteria except on criterion g_i on which its evaluation is increased by a significant amount (but not exceeding the veto threshold) relative to the scale of g_i, (b_i P_i b_0).

---

[3] Each upper or lower bound comes either from an original inequality, or from two "parent bounds". It is then easy to retrieve the "ancestry chain" of a bound so as to identify its origins.

$b_{i,j}$: Alternatives indentical to b_0 on all criteria except g_i and g_j on which its evaluation is increased by a significant amount (but not exceeding the veto threshold) relative to the scales of g_i and g_j, ($b_{i,j}$ P_i b_0 and $b_{i,j}$ P_j b_0).

b_J: Alternatives identical to b_0 on all criteria except on criteria contained in the coalition J ($J \subseteq F$) on which its evaluation is increased by a significant amount (but not exceeding the veto threshold) relative to the scales of the considered criterion (b_J P_i b_0, $\forall i \in J$).

If the DM and the analyst agree to ground decision aid on an ELECTRE type MCAP, then the knowledge of the comprehesive preference situation between two fictitious alternatives allows us to infer information on importance parameters of this MCAP. In fact, when the DM states b_{J_1} P b_{J_2}, it means that the advantages on criteria contained in J_1 loom larger than the ones on criteria contained in J_2, i.e., the coalition J_1 is more important than the coalition J_2 ($J_1 \gg J_2$). As each assertion of the relation $\gg$ (*more important than between disjoint coalitions of criteria*) is formalized, in the considered MCAP, by $\forall J_1, J_2 \subseteq F \quad J_1 \succ J_2 \quad \Leftrightarrow \quad \sum_{i \in J_1} k_i > \sum_{i \in J_2} k_i$, we infer from each preference relation between b_{J_1} and b_{J_2} an inequality on k_js.

Similarly, an indifference between b_{J_1} and b_{J_2} leads to stating an equality on k_js: $\sum_{i \in J_1} k_i = \sum_{i \in J_2} k_i$. So as to soften the highly reductive effect of such an equality on the set of admissible values for k_js, we propose to treat indifferences in the following way:

$$b_{J_1} I b_{J_2} \quad \Leftrightarrow \quad \left|\sum_{j \in J_1} k_j - \sum_{j \in J_2} k_j\right| < k_{min} \quad \text{with} \quad k_{min} = \underset{j \in F}{Min}\{k_j\}$$

The "strength" of the indifference is then reduced to an equality on k_js "to within k_{min}".

Moreover, it is important to stress what underlies the technique of saturated segmented description. When the polyhedron of admissible values for k_js becomes empty, this means that it is not possible to find values for k_js compatible, in the considered model, with the assertions expressed by the DM.

3.3.2. Preliminary step: eliciting discrimination thresholds.

The role of indifference and preference thresholds q_j and p_j is to specify the preferences restricted to the significance axis of a criterion; they do not refer directly to the notion of RIC. However these thresholds interact with the inter-criteria preferential parameters and may have an indirect influence on the role of each criterion in the aggregation. Moreover, the definition of b_J requires the knowledge of p_j. So as to determine values for these thresholds, the analyst may refer to the ill-determined nature of some constituent elements of criteria (see [Roy 85], chap. 9 and [Roy et al. 86]). In certain situations, these thresholds may be elicited through an interaction with the DM.

In the implementation of the ELECTRE methods (see [Vallee & Zielniewicz 94]), p_j and q_j are defined as afine fonctions of evaluations, i.e., such that:

$$\begin{cases} q_j(g_j(a)) = a_q \cdot g_j(a) + b_q \\ p_j(g_j(a)) = a_p \cdot g_j(a) + b_p \end{cases}$$

So as to obtain such an information, we propose to proceed in the following way:

- Determine a neutral evaluation n_j (neither attractive, nor repulsive, neither representing an avantage, nor a drawback on the considered criterion).
- Determine an attractive evaluation a_j representing a significant advantage with regard to the neutral level.

Given a_j and n_j[4], we elicit $q_j(n_j)$, $p_j(n_j)$, $q_j(a_j)$ and $p_j(a_j)$[5]. So as to minimize the duration of the interaction, questions are asked following a dichotomic search. The following algorithm determines $q_j(n_j)$ and $p_j(n_j)$ ($q_j(a_j)$ and $p_j(a_j)$ are obtained similarily).

```
For each criterion j do
|       min ← n_j
|       max ← a_j
|       While max-min > ε                     (ε: proportion of the range of the scale)
|       Do
|       |       If max+min/2 I_j n_j
|       |       |       Then min ← max+min/2
|       |       |       Else max ← max+min/2
|       |       Endif
|       End
|       q_j(n_j) ← max+min/2
|       min ← q_j(n_j), max ← a_j
|       While max-min > ε
|       Do
|       |       If max+min/2 P_j n_j
|       |       |       Then max ← max+min/2
|       |       |       Else min ← max+min/2
|       |       Endif
|       End
|       p_j(n_j) ← max+min/2
End
```

Determining the discrimination thresholds in this way supposes that:

- criteria are evaluated on a continuous scale (or discrete with a large number of levels),
- criteria are not the result of a sub-aggregation,
- the DM knows precisely how evaluations are determined.

---

[4] The analyst should check that a_j and n_j are such that $a_j P_j n_j$.

[5] It is possible to determine more points so as to make a linear regression, but it requires more questions.

3.3.3. Step 1: Rank criteria by order of importance

The first step in the questioning procedure consists of searching for a pre-order on the k_js. To achieve this, the alternatives $b_1,b_2,\ldots\ldots,b_n$ are presented to the DM; he should then determine the alternative b_h he considers the best. Then, it holds $b_hPb_i\ \forall i\neq h$; hence $k_h>k_i\ \forall i\neq h$. When several alternatives $b_{h_1},b_{h_2},\ldots\ldots,b_{h_p}$ are judged to be the best and indifferent to one another, we have: $\forall h\in H,\ \forall h'\in F\backslash H\ \ b_hPb_{h'}$ and $\forall h,h'\in H\ \ b_hIb_{h'}$ with $H=\{h_1,h_2,\ldots\ldots,h_p\}$. Then $k_{h_1}=k_{h_2}=\ldots\ldots=k_{h_p}>k_{h'}\ \forall h'\in F\backslash H$.

The best alternative(s) is (are) then deleted from the initial list and the DM must choose the best one in the list of remaining alternatives, etc. We then obtain a pre-order on the k_js and it is always possible to renumber the criteria so that: $k_1\leq k_2\leq\ldots\ldots\leq k_n$. When several criteria are of equal importance, we keep only one representative of the equivalence class, during the rest of the questioning procedure, so as to obtain, after a second renumbering: $k_1<k_2<\ldots\ldots<k_{n'}$.

3.3.4. Step 2: determining groups of criteria which are close in importance

The second step aims at partitioning the set of criteria (ranked by importance beforehand) according to a specific condition that can be interpreted as defining a partition into groups of "relatively close" criteria when considering their relative importance. Each group of criteria G_i is defined by the index h(i) of the least important criterion of the group. The p groups are such that:

$$\left[\begin{array}{l}\{g_1,g_2,\ldots,g_{h(1)+i},\ldots,g_{h(2)-1}\}\\ \{g_{h(2)},g_{h(2)+1},\ldots,g_{h(2)+i},\ldots,g_{h(3)-1}\}\\ \qquad\vdots\\ \{g_{h(p)},g_{h(p)+1},\ldots,g_{h(p)+i},\ldots,g_n\}\end{array}\right.$$

The h(i) are defined by: $h(i+1)=\min\limits_{j\in F}\{j/g_j\succ\{g_{h(i)},g_{h(i)+1}\}\}$ with $h(1)=1$. The groups must be interpreted as: $g_{h(i+1)}$ is the least important criterion that remain more important than the coalition $\{g_{h(i)},g_{h(i)+1}\}$. The information necessary for determining the partition is the following:

$$\left[\begin{array}{l}g_{h(i+1)}\succ\{g_{h(i)},g_{h(i)+1}\}\\ \text{non}\left[g_{h(i+1)-1}\succ\{g_{h(i)},g_{h(i)+1}\}\right]\end{array}\right.\quad\forall i=\{1,2,\ldots\ldots,p\}$$

In order to obtain this information, the DM must respond to pairwise comparisons of alternatives whose evaluations vary on 3 criteria. The protocole is the following:

- Do you prefer $b_{1,2}$ or b_n ?
  - if the answer is $b_{1,2}$ P b_n then $k_1+k_2>k_n$, we stop here for this step.
  - if the answer is $b_{1,2}$ I b_n then $|k_n-k_1-k_2| < k_1$, we stop here for this step.
  - if the answer is b_n P $b_{1,2}$ then we ask the following question.
- Do you prefer $b_{1,2}$ or b_{n-1} ?
  - if the answer is $b_{1,2}$ P b_{n-1} then $k_1+k_2>k_{n-1}$, we stop here for this step.
  - if the answer is $b_{1,2}$ I b_{n-1} then $|k_{n-1}-k_1-k_2|<k_1$, we stop here for this step.
  - if the answer is b_{n-1} P $b_{1,2}$ then we go on until we find the smallest h(2) such that not[$b_{h(2)-1}$ P $b_{1,2}$].

We then continue by asking the question:

- Do you prefer $b_{h(2),h(2)+1}$ or b_n
  - if the answer is $b_{h(2),h(2)+1}$ P b_n then $k_{h(2)}+k_{h(2)+1}>k_n$, we stop here for this step.
  - if the answer is $b_{h(2),h(2)+1}$ I b_n then $|k_n-k_{h(2)}-k_{h(2)+1}|<k_1$, we stop here for this step.
  - if the answer is b_n P $b_{h(2),h(2)+1}$ then we go on until we find the smallest h(3) such that not[$b_{h(3)-1}$ P $b_{h(2),h(2)+1}$].
- and so on until the whole familly of criteria is partitioned.

The second step of the questioning procedure is summarized in the following algorithm:

```
i←1
h(i)←1
While h(i)+2 ≤ number-of-crit
Do
    j←number-of-crit
    Repeat
        Compare b_j to b_h(i),h(i)+1
        j←j-1
    Until not[b_j P b_h(i),h(i)+1] or h[i]+2>j
    If b_h(i),h(i)+1 P b_j
        Then h(i+1)←j+2
        Else h(i+1)←h(i-1)+2
    Endif
    i←i+1
End
number-of-group ← i-1
```

So as to obtain the required information, this algorithm contains linear sequences of questions. It is possible to reduce significantly the number of questions by applying a dichotomic segmentation rule: instead of comparing $b_{h(i),h(i)+1}$ to b_n, b_{n-1}, etc. successively, it is more efficient to determine h(i+1) using a dichotomic search in the interval [h(i)+2,n].

3.3.5. Step 3: Evaluating of the "distance" between groups of criteria

At the end of the first two steps, we obtain a partition of the set of criteria (subsequently ordered) in groups of "relatively close" criteria, from the point of view of their relative importance. The third step aims at evaluating the "distance" between these groups. The procedure seeks, for each group, the coalition composed of the two least important criteria which is more important than the least important criterion of the group just above it. More precisely, for each group G_i we search for $m_1(i)$ and $m_2(i)$ such that:

$$\begin{bmatrix} \{g_{m_1(i)}, g_{m_2(i)}\} \succ g_{h(i-1)} \\ \forall a,b \in F \qquad \begin{array}{c} a \leq m_1(i) \quad \text{et} \quad b \leq m_2(i) \\ \text{(one inequality at least being strict)} \end{array} \end{bmatrix} \Rightarrow g_{h(i-1)} \succ \{g_a, g_b\}$$

In order to find $m_1(i)$ and $m_2(i)$, the sequence of questions and answers is the following:

- Do you prefer $b_{h(i)+1,h(i)+3}$ or $b_{h(i+1)}$?
  - If the answer is $b_{h(i)+1,h(i)+3}$ P $b_{h(i+1)}$, then $k_{h(i)+1}+k_{h(i)+3}>k_{h(i+1)}$, we stop for this group and we go on to the next group.
  - If the answer is $b_{h(i)+1,h(i)+3}$ I $b_{h(i+1)}$, then $|k_{h(i)+1}+k_{h(i)+3}-k_{h(i+1)}|<k_1$, we stop for this group and we go on to the next group.
  - If the answer is $b_{h(i+1)}$ P $b_{h(i)+1,h(i)+3}$, then we ask the following question:

- Do you prefer $b_{h(i)+1,h(i)+4}$ or $b_{h(i+1)}$?
  - If the answer is $b_{h(i)+1,h(i)+4}$ P $b_{h(i+1)}$, then $k_{h(i)+1}+k_{h(i)+4}>k_{h(i+1)}$, we stop for this group and we go on to the next group.
  - If the answer is $b_{h(i)+1,h(i)+4}$ I $b_{h(i+1)}$, then $|k_{h(i)+1}+k_{h(i)+4}-k_{h(i+1)}|<k_1$, we stop for this group and we go on to the next group.
  - If the answer is $b_{h(i+1)}$ P $b_{h(i)+1,h(i)+4}$, then we go on until we find λ such that $b_{h(i)+1,h(i)+\lambda}$ P $b_{h(i+1)}$. If it is not possible to find such a λ, then we ask the following question :

- Do you prefer $b_{h(i)+2,h(i)+3}$ or $b_{h(i+1)}$?
  - If the answer is $b_{h(i)+2,h(i)+3}$ P $b_{h(i+1)}$, then $k_{h(i)+2}+k_{h(i)+3}>k_{h(i+1)}$, we stop for this group and we go on to the next group.
  - If the answer is $b_{h(i+1)}$ I $b_{h(i)+2,h(i)+3}$, then $|k_{h(i)+2}+k_{h(i)+3}-k_{h(i+1)}|<k_1$, we stop for this group and we go on to the next group.
  - If the answer is $b_{h(i+1)}$ P $b_{h(i)+2,h(i)+3}$, then we ask the following question:

- Do you prefer $b_{h(i)+2,h(i)+4}$ or $b_{h(i+1)}$?
  - If the answer is $b_{h(i)+2,h(i)+4}$ P $b_{h(i+1)}$, then $k_{h(i)+2}+k_{h(i)+4}>k_{h(i+1)}$, we stop for this group and we go on to the next group.
  - If the answer is $b_{h(i)+2,h(i)+4}$ I $b_{h(i+1)}$, then $|k_{h(i)+2}+k_{h(i)+4}-k_{h(i+1)}|<k_1$, we stop for this group and we go on to the next group.
  - If the answer is $b_{h(i+1)}$ P $b_{h(i)+2,h(i)+4}$, then we continue until we find λ_1 and λ_2 such that not[$b_{h(i+1)}$ P $b_{h(i)+\lambda_1,h(i)+\lambda_2}$].

In the third step, the sequence of questions follows the algorithm given below:

```
For each group G_i
Do
    m_1(i) ← h(i)
    m_2(i) ← h(i)+1
    stop ← false
    Repeat
        If m_2(i)+1 < h(i+1)
            Then m_2(i) ← m_2(i)+1
            Else If m_1(i)+2 < h(i+1)
                Then m_1(i) ← m_1(i)+1
                     m_2(i) ← m_1(i)+1
                Else stop ← true
            Endif
        Endif
        If not stop
            Then Compare b_h(i+1) to b_m1(i),m2(i)
        Endif
    Until stop or not[b_h(i+1) P b_m1(i),m2(i)]
End
```

3.3.6. Step 4: obtaining information so that each k_j has an upper bound

At the end of the third step, each k_j has a lower bound (at least k_{j-1}), but it does not necessarily have an upper bound. This means that the domain of admissible values for $k=(k_1,k_2,.....,k_n)$ is open. This step aims at providing an upper bound for each k_j. A sufficient condition for the polyhedron to be closed is that k_n has an upper bound and so has k_n/k_1. Indeed we may deduce from $k_n/k_1<\alpha$ that $k_j<\alpha.k_1$ $(\forall j)$. So as to bound this ratio, the analyst may ask the following question: suppose that we can imagine α criteria having the same importance as g_1; let us consider the fictitious alternative b_1^α whose evaluations are average (i.e., identical to b_0) except on these α criteria on which its evaluation is improved by a significant amount. Formally, the comparaison of b_1^α and b_n (α being variable) allows us to obtain an upper bound for the ratio k_n/k_1. This technique puts only few questions to the DM; however, the complexity of the required interaction can lead to an uncertain and imprecise answer.

Another technique may be used. We consider $J\subseteq F$ the set of criteria g_j for which k_j has no upper bound at the end of step 3 and such that there exists a criterion g_i that is less important than g_j and possesses an upper bound. The fourth step is defined by the following sequence of questions:

```
For each segment j ∈ J
Do
|       Let kr+ks be an upper bound of a segment lower than j and possessing an upper
|       bound (we suppose kr<ks)
|       stop ← false
|       Repeat
|       |       Compare bj to br,s
|       |       If s+1 < j
|       |       |   Then s ← s+1
|       |       |   Else If r+2 < j
|       |       |           |   Then r←r+1
|       |       |           |            s←r+1
|       |       |           |   Else stop ← true
|       |       |           Endif
|       |       Endif
|       Until  stop  or  br,s P bj
End
```

An advantage of this technique lies in the fact that its interaction deals with pairwise comparisons of alternatives (which is consistent with the rest of the procedure). However, this algorithm does not always provide an upper bound to all k_j; the coefficients of some of the least important criteria may remain unbounded (this leaves the polyhedron open). For instance, it is difficult to find an upper bound for k_3 when $k_3>k_1+k_2$.

3.3.7. Step 5: adding supplementary inequalities so as to reduce the polyhedron of admissible values

The fifth step aims at reducing the polydron of admissible values for k. This step is necessary when the obtained polyhedron leads to considering a very large number of sets of importance coefficients to be valid.

The questions deal with two alternatives ($b_{i,j}$ et $b_{k,l}$) whose evaluations vary on 4 criteria. In order for the answers to provide supplementary information, it is necessary that $k_i<k_k,k_l<k_j$. The choice of these 4 criteria (verifying $k_i<k_k,k_l<k_j$) must be made with regard to the polyhedron obtained. It is not necessary, therefore, to automate this step (the choice of these questions is left to the analyst). It should be noted that inconsistencies usually appear at this stage (see 3.2.2 how to reduce such inconsistencies).

3.3.8. Determining several admissible weight vectors

When the polyhedron of admissible values for k is closed, an interval of variation for each k_j can easily be inferred from the saturated segmented description: each k_j lies in a segment whose bounds are functions of the k_i verifying $k_i<k_j$. These intervals are the following:

$$m_2(k_1) < k_2 < M_2(k_1)$$
$$m_3(k_1,m_2(k_1)) < k_3 < M_3(k_1,M_2(k_1))$$
$$m_4(k_1,m_2(k_1),m_3(k_1,m_2(k_1))) < k_4 < M_4(k_1,M_2(k_1),M_3(k_1,M_2(k_1)))$$
$$\vdots$$
etc.

The value of the lower and upper bound of each coefficient k_j is obtained by assigning to the k_i (such that $i<j$) in the function $m_j(k_1,k_2,.....,k_{j-1})$ and $M_j(k_1,k_2,.....,k_{j-1})$ their minimum or maximum value, respectively, according to the sign of the coefficient in the linear forms $m_j(k_1,k_2,.....,k_{j-1})$ and $M_j(k_1,k_2,.....,k_{j-1})$ respectively. It should be noted that these extremes values for k_j, can be reached only for particular values of k_i such that $k_i<k_j$.

Moreover, the information obtained enables us to build easily a large number of admissible vectors $k=(k_1,k_2,...,k_n)$. A "central" weight vector may be generated by assigning the central value of its own variation interval to each k_j. When we bring the value of each k_j closer to the upper bound (or to the lower bound), we obtain vectors that are "wider" (or "narrower").

3.3.9. Elicitation of the veto thresholds

Like the coefficients k_j, the veto thresholds $v_j(g_j)$ are inter-criteria preferential parameters. They account for an aspect of the notion of RIC which is distinct from that modelled by the coefficients k_j[6]. We should recall that the veto thresholds aim at impoverishing the outranking relation through the invalidation of assertions aSb that satisfy the concordance condition.

Before trying to elicit thresholds $v_j(g_j)$, it is important to determine whether the DM wants to attach some veto power to the criteria. Moreover, the analyst must verify whether this veto power may become effective in comparing alternatives, i.e., if $v_j(g_j)<L_j$ with $L_j = \max_{a\in A}(g_j(a)) - \min_{a\in A}(g_j(a))$. When the veto has a significant effect on the result, further interaction with the DM is necessary. The software implementation of the ELECTRE methods considers the thresholds $v_j(g_j)$ as afine functions of evaluations. Several interaction modes have been proposed so as to determine values for $v_j(g_j)$ with the DM.

It is possible to take advantage of the meaning that the DM gives to the criteria and the role he wants them to play. [Roy & Bouyssou 93] (chap. 8 et 9) propose using

---

[6] These two facets of the importance of criteria are usually linked; however, it is formally possible for a criterion to have simultaneously a low weight and a large veto power.

the ratio v_j/p_j. We will propose below a method for determining an interval of variation for v_j. We will determine values for $v_j(n_j)$ and $v_j(a_j)$ (n_j and a_j being neutral and attractive evaluations respectively).

So as to determine these values, we proceed indirectly by asking questions concerning fictitious alternatives. We denote $b^n=(n_1,n_2,.....,n_n)$ and $b^a=(a_1,a_2,.....,a_n)$; let b^n_J and b^a_J be the alternatives having the same evaluations as b^n and b^a respectively except on the criteria $j\in J$ ($J\subseteq F$) on which its evaluation is increased by p_j. We denote b^n_{j+x} and b^a_{j+x} the alternatives having the same evaluations as b^n and b^a respectively except on criterion j on which their evaluations are increased by x. So as to bound $v_j(n_j)$ and $v_j(a_j)$, the DM should compare b^n_{j+x} to b^n_J and b^a_{j+x} to b^a_J, x being variable and $J=F\backslash\{j\}$ (in what follows, we will explain only how to determine $v_j(n_j)$; we will proceed similarily with $v_j(a_j)$).

It should be noted that the assertion b^n_J S b^n_{j+x} satisfies the concordance condition[7]. In the comparaison of b^n_J and b^n_{j+x}, the DM may either prefer b^n_J or hesitate because of difficulties when comparing these alternatives. In the first case, x constitutes a lower bound for $v_j(n_j)$ while in the second case, it specifies an upper bound for $v_j(n_j)$. So as to bound $v_j(n_j)$ we proceed through a dichotomic segmentation of the interval $[p_j,L_j]$ (L_j being the range of the scale); this leads to the following algorithm:

```
lower ← p_j
upper ← L_j
While upper-lower > ε     (ε being a proportion of L_j-p_j)
Do
    x ← upper+lower/2
    Compare b^n_J and b^n_{j+x}
    If b^n_J P b^n_{j+x}
        Then lower ← upper+lower/2
        Else upper ← upper+lower/2
    Endif
End
```

Determining intervals for $v_j(n_j)$ and $v_j(a_j)$ allows us to define the coefficients of two lower and upper bound fonctions $v^m_j(g_j)$ and $v^M_j(g_j)$ (we should verify that these fonctions are such that $v^m_j(g_j(a))<v^M_j(g_j(a))$ $\forall a\in A$). The remarks concerning the conditions for using the algorithm for determining discrimination thresholds (see 3.3.2) also apply for the present algorithm.

---

[7] except in very particular cases in which de k_j is very large $(k_j>\sum_{i\neq j}k_i)$.

3.4. Implementation of the method

3.4.1. Role of such a tool in a decision aid process

DIVAPIME is to be inserted in a decision aid process in which a multiple criteria model is used. The elicitation of preferential information is a crucial phase of the modelisation that is often problematical. The proposed software aims at supporting this elicitation step (when an ELECTRE type MCAP is chosen to model the decision process). Figure 2 places DIVAPIME in a classical multicriteria decision aid process.

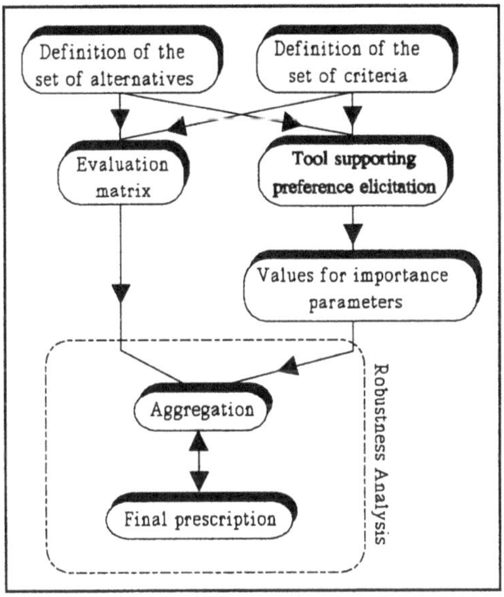

Figure 2

The way this tools should be inserted in a decision process must be analysed. Several studies concerning the elicitation of the RIC have shown that there is a stumbling stock for any ETIP, namely the existence of a gap between the information underlying the answers given to the analyst by the DM, and the way these answers are interpreted in the model. We believe that the presence of an analyst may partially avoid this problem as he/she can verify whether the DM's understanding and interpretation of the questions is correct.

Hence, this software is essentially intended to the analysts and aims at facili-tating their work while enhancing the user-friendliness of the interaction with the DM. Although this tool has some of the characteristics of a decision support system (DSS)[8], it is not, in its present form, intended for the DM. Its role is not necessarily to transfer some of the activities of the analyst to the DM. It was not designed as a substitute for the analyst, whose role remains essential: the latter must explain to the DM the nature and the meaning of each of the preferential parameters as well as the way in which the DM's answers will be interpreted.

However, an independant use of this software by a DM with good background knowledge of the ELECTRE methods is possible. The proposed tool may then be viewed as a DSS that helps the DM in formalizing his preferences and eliciting the preferential

---

[8] Assisting the DM in ill-structured tasks, helping rather than replace the DM's judgment, using the interactive possibilities of the tools (see [Sprague & Carlson 82], [Levine & Pomerol 89]).

parameters required by the ELECTRE methods. On the other hand, it may be dangerous to put such a tool in the hands of a novice in MCDA, since it would be difficult to know what the basis of the output obtained would be[9].

3.4.2. Presentation of the DIVAPIME software: general description

DIVAPIME software (**D**étermination d'**I**ntervalles de **VA**riation des **P**aramètres d'**I**mportance des **M**éthodes **E**lectre) works with MS-DOS 3.2 or higher on a IBM PC with

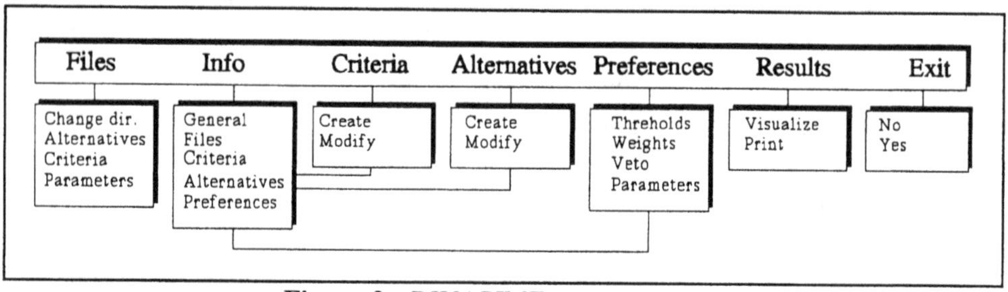

Figure 3 : DIVAPIME menu structure

a minimum of 640 Kb memory and a VGA color monitor. It is implemented with Borland Turbo Pascal. The different options proposed are described in figure 3.

The content of the options in the main menu is as follows:

Files: provide the list of criteria files, alternatives files and parameter set files.

Info: provides general information on the ETIP (in particular the way answers are used) and explanations concerning each option. This option may be activated from any option as an on-line help feature.

Criteria: input, loading and/or modification of a set of criteria.

Alternatives: input, loading and/or modification of a set of alternatives.

Preferences: Evaluation of preferential parameters through a questioning procedure.

Results: display or print obtained intervals of variation for preferential parameters.

The standard scheme of a DIVAPIME session is the following:

- Input (or loading/modification) of a set of criteria.
- Input (or loading/modification) of the set of alternatives[10].
- Determination of the discrimination thresholds p_j and q_j.

---

[9] However, DMs frequently lead real world decision aid processes without the help of an analyst. The use of DIVAPIME in such a context will in any case be less controversial than the most frequently used method, i.e., direct numerical evaluation of the parameters.

[10] This option is included in the software because the DM must have a precise perception of the set of alternatives so as to express judgments on the importance of criteria.

- Determination of intervals of variation for the importance coefficients k_j and possibly for the veto thresholds v_j[11].
- Generation of one or several sets of preferential parameters.

The interested reader will find in [Mousseau 93] an illustrative example of how DIVAPIME may be used in a decision aid process. It should also be mentioned that a new Windows version of this software will be implemented and will be adapted to a multi-actor situation.

3.5. Extensions

3.5.1. Adapting the method to the problem formulation

We believe that it is erroneous to conceive an ETIP without taking into account the problem formulation adopted in the modelling of the decision problem: the questioning procedure should correspond to the way the DM analyses the problem (Pα: choice, Pβ: assignment or Pγ: ranking, see [Roy 85] and [Bana e Costa 93]). In our presentation, the proposed method is mainly adapted to a problem formulated in ranking terms. The questions put to the DM should be modified so as to become appropriate for the choice and assignment problem formulation (Pα and Pβ).

In the case of a choice problem, the answer to a pairwise comparison of alternatives could be "*I do not want to choose either of them*". Such an answer is hardly interpretable and appears when both alternatives are judged to be insufficiently attractive in order to be selected in the final prescription. It is then possible to "force" the DM's answer by puting him/her in a situation in which two alternatives are the only available options. Another way to avoid such a problem consists in proposing comparisons to the DM, in which both alternatives are "attractive". In the method proposed here, this may be done by changing the definition of the reference alternative b_0 (see 3.3.1) whose evaluations on all criteria should be attractive (rather than neutral).

When the decision situation is modelled through an assignment problem formulation, questioning the DM on the basis of pairwise comparisons of alternatives poses a more fundamental problem: the assignment of alternatives to a category is not founded on the comparisons of alternatives but on an absolute evaluation. In some cases, pairwise comparison of alternatives may be used. However, this mode of interaction is usually unsuitable and should be modified for the dialogue between the analyst and the DM to conform to the logic of Pβ.

---

[11] There is no constraint of precedence between the evaluation of k_js and v_js.

One way to proceed is to ask the questions in terms of an assignment as follows. We denote $\{K_1,K_2,\ldots,K_p\}$ the ordered set of predefined categories ($i>j \Leftrightarrow$ categorie K_i is better than categorie K_j). b_0 is a fictitious alternative conceived of in such a way that b_0 is assigned to K_i (we note $b_0 \rightarrow K_i$). b_{J_1} and b_{J_2} ($J_1 \subseteq F$, $J_2 \subseteq F$ and $J_1 \cap J_2 = \emptyset$) are two fictitious alternatives defined as in 3.3.1. If the DM assigns b_{J_1} and b_{J_2} to the categories K_{λ_1} and K_{λ_2} respectively, then it holds: $\lambda_1 > \lambda_2 \Rightarrow J_1 \gg J_2$. From this information, we can infer values (or intervals of variation) for importance parameters through the MCAP used[12].

3.5.2. Extension to a multiple DM framework

DIVAPIME is presently intented to be used in a decision situation in which a single DM is involved. Several ways can be considered to extend this ETIP (and its implementation) to decision situations in which several DMs interact[13].

The first of these is to force the DMs to answer collectively the questions asked during the procedure. From this perspective, the multi-actor aspect of the problem is not directly managed by the method but is taken over by the DMs who must discuss their arguments before answering each question. Hence, no adaptation of the method is required[14]. This approach is particularily suitable when criteria represent viewpoints of specific actors. In this case, it is difficult for some of these actors to answer all questions as they may have a precise opinion on the importance of criteria only with regard to a subset of criteria. During a collective questioning procedure, it may occur that several DMs disagree on the answer to one (or more) question(s). It is then advisable to put aside temporarily the inequalities corresponding to these questions and to reintegrate them at the end in order to generate the corresponding polyhedrons.

---

[12] This approach, although more suitable for the assignment problem formulation, poses problems when the number of categories is low.

[13] However such an extention is conceivable only in consensual multi-actor decision situations. In fact, in a decision problem in which opposition between actors is acute, it will be difficult for DMs to reach an agreement concerning the values of preferential parameters; seeking such values will only underline points of conflict but will not be able to orient the decision process towards a compromise solution.

[14] However, each DM should form his/her own opinion concerning the importance of criteria before the general discussion: it is possible, for example, to carry out all (or a part of) the questioning procedure with each DM individually.

A second approach consists of determining intervals of variation for importance parameters with each DM individually and trying to group DMs whose opinions are "close". The simplest situation occurs when the intersection of all polyhedrons of admissible values for k_j is non-empty; a vector $k=(k_1,k_2,.....k_n)$ can then be chosen in the polyhedron defined by this intersection. Nevertheless, this polyhedron is frequently empty and such a vector cannot be found. In this case, another way to proceed would be to take advantage of automatic classification techniques so as to constitute a partition of the set of DMs. It is then simple to generate a vector k representing each group of DMs. If the prescriptions stemming from the different sets of parameters converge, then it is not necessary to reduce divergences of opinion between DMs. In the opposite case, a discussion between DMs is imperative to reach a compromise; this discussion should not revolve directly around values for parameters but rather on the sequence of questions and answers for which a divergence has appeared.

Conclusion

A careful analysis of the notion of Relative Importance of Criteria (RIC) proves this notion to be more complex than is commonly assumed. We have shown (section 1) that the information that underlies this notion is much richer than that contained in the importance parameters used in the various multicriteria models. Hence, these parameters constitute a simplistic way of taking RIC into account. Moreover, the meaning of such parameters varies across models.

These considerations, together with an empirical analysis of how DMs understand the notion of RIC, constitute a basis on which Elicitation Techniques for Importance Parameters (ETIP) may be developed (section 2). In addition, the role of ETIP will differ according to whether or not preferences are assumed to exist prior to the modeling process. If we assume that preferences pre-exist, ETIP aims at estimating pre-existing information; if we assume they do not pre-exist, ETIP only provides parameters consistent (according to an aggregation rule) with some assertions stated by the DM.

Section 3 is devoted to the presentation of an ETIP adapted for the ELECTRE methods. The interaction with the DM proceeds by means of pairwise comparisons of fictitious alternatives. This technique tests the consistency of the DM's answers with the aggregation rule used and provides, as output, an interval of variation for each parameter. This output constitutes an interesting starting point for robustness analysis.

References

Bana e Costa C. (1993) "*Les problématiques dans le cadre de l'activité d'aide à la décision*", Document n°80, LAMSADE, Université Paris-Dauphine.

Beattie J. & Baron J. (1991) "*Investigating the effect of stimulus range on attribute weight*", Journal of Experimental Psychology: Human Perception and Performance, 17(2), 571-585.

Brans J.P., Mareschal B. & Vincke Ph. (1984) "*PROMÉTHÉE: A new family of outranking methods in multicriteria analysis*" in J.P. Brans (ed.), Operational Research '84, Elsevier Science Publishers B.V. (North-Holland), 408-421.

Fishhoff B., Slovic P. & Lichtenstein S. (1989) : "*Knowing what you want: Measuring labile values*" in Bell-Raiffa-Tversky (eds.), Decision Making: Descriptive, Normative and Prescriptive Interactions, Cambridge University Press, 398-421.

Keeney R.L. & Raiffa H. (1976), (2nd ed. 1993) "*Decision with multiple objectives: preferences and value tradeoffs*", John Wiley & Sons.

Levine P. & Pomerol J.Ch. (1989) "*Systèmes interactifs d'aide à la décision et systèmes experts*", Hermès.

Mousseau V. (1992) "*Analyse et classification de la littérature traitant de l'importance relative des critères en aide multicritère à la décision*", RAIRO / Operations Research, 26 (4), 367-389.

Mousseau V. (1993) "*Problèmes liés à l'évaluation de l'importance relative des critères en aide multicritère à la décision : réflexions théoriques, expérimentations et implémentations informatiques*", Thèse de doctorat, Université Paris-Dauphine.

Paynes J.W., Bettman J.R. & Johnson J.E. (1993) "*The adaptative decision making*", Cambridge University Press.

Paynes J.W., Bettman J.R. & Johnson J.E. (1992) "*Behavioral decision research: a constructive processing perspective*", Annual Review of Psychlogy, 43, 87-131.

Podinovskii V.V. (1988) "*Criteria importance theory*" in Lewandowski, Volkovich (Eds.) "Multiobjective problems of mathematical programming", Proceedings, Yalta, USSR, Springer-Verlag.

Podinovskii V.V. (1994) "*Criteria importance theory*", Mathematical Social Sciences, 27, 237-252.

Roubens M. (1982) "*Preference relations on actions and criteria in multicriteria decision making*", EJOR, 10, 51-55.

Roy B. (1970) "*Algèbre moderne et théorie des graphes orientées vers les sciences économiques et sociales*", Tome 2, Dunod.

Roy B. (1985) "*Méthodologie multicritère d'aide à la décision*", Economica.

Roy B. (1991) "*The outranking approach and the foundations of ELECTRE methods*", Theory and Decision, 31 (1), 49-73.

Roy B. (1993) "*Decision science or decision-aid science*", EJOR, 66 (2), 184-203.

Roy B. & Bouyssou D. (1993) "*Aide multicritère à la décision : Méthodes et cas*", Economica.

Roy B. & Mousseau V. (1995) "*A theoretical framework for analysing the notion of relative importance of criteria*", Research Paper, LAMSADE, Université Paris-Dauphine.

Roy B., Présent M. & Silhol D. (1986) "*A programming method for determining which Paris metro stations should be renovated*", EJOR , vol. 24, pp. 318-334.

Saaty T.L. (1980) "*The Analytic Hierarchy Process*", Mc Graw Hill, New York.

Sprague R.H. & Carlson E.D. (1982) "*Building effective decision support systems*", Prentice Hall.

Vallée D. & Zielniewicz P. (1994) "*Electre III/IV version 3.x : Aspects méthodologiques*", Document du Lamsade n°85, Université Paris-Dauphine.

Vansnick J-Cl. (1986) "*On the problem of weight in multiple criteria decision making (the non compensatory approach)*", EJOR, 24, 288-294.

Vincke Ph. (1990) "*Basic concepts of preference modelling*", in Bana e Costa (ed.) "Readings in multiple criteria decision aid", Springer Verlag, 101-118.

Weber M. & Borcherding K. (1993) "*Behavioral influences on weight judgments in multiattribute decision making*", EJOR, 67(1), 1-12.

II. PROBLEM STRUCTURING

UNSTRUCTURED PROBLEMS AND DEVELOPMENT OF PRESCRIPTIVE DECISION MAKING METHODS

O. I. Larichev and H.M. Moskovich

Institute for System Analysis

Russian Academy of Sciences (Moscow)

Abstract

A class of unstructured decision making problems is under consideration. Unstructured problems are the problems with the majority of qualitative parameters with unknown quantitative dependencies. The peculiarities of these tasks are discussed, and requirements for decision aid tools are formulated: psychologically valid measurements and elicitation procedures, constistency testing, and possibility to communicate the result. Method ZAPROS is described as an example of the decision aid, meeting these requirements. Peculiarities of the method are illustrated on an example.

KEYWORDS: multiattribute, decision aid, ZAPROS, prescriptive methods

---

The work is partly supported by the grant 93-012-442 of the Russian Basic Research Fund (FBRF), and partly by the grant DPP9213392 of the National Science Foundation (NSF).

P.M. Pardalos et al. (eds.), Advances in Multicriteria Analysis, 47–80.

INTRODUCTION

According to Simon and Newell (1958) decision problems may be divided into three main groups: well-structured problems, ill-structured problems and unstructured problems.

Problems in which dependencies between the parameters are known and may be expressed in a formal way, are considered to be well-structured problems. Problems of this class are being rather successfully solved by operations research methods. Tasks of linear and dynamic programming, optimal control, and others are typical representatives of this class.

Ill-structured problems may be considered to be of intermediate nature as they deal with qualitative as well as quantitative parameters, but qualitative, unknown and undefined problem elements tend to dominate in these tasks. The discrepancy between these two classes of decision making problems reflects the discrepancy between operations research and decision making theory. The second class acquire rather diversified set of problems, which themselves may be divided into three subclasses according to the proportion of subjective and objective elements in them.

Ill-structured problems for which there exists some objective model (analogous to the models of operations research), but the quality of decision is evaluated by multiple criteria, forms the first subclass. Multicriteria analogies to the classical tasks of operations research may be considered as examples of such problems: multicriteria mathematical programming (Steuer, 1986), multicriteria assignment problem (Larichev & Kozuharov, 1979), multicriteria bin packing (Larichev & Furems, 1987), and others. In these problems phases of analysis, carried out by a decision maker are combined with phases of calculations on computers. As B.Roy (1985) marked, this approach consists of preference elicitation simultaneously with investigation of the feasible set of actions to find out the effective solutions.

Second subclass may be formed by the problems with qualitative and quantitative parameters, conditionally divided into two groups: those characterizing cost, and those characterizing benefit. Many problems connected with military tasks for which the system analysis approach (Quade, 1984) has been developed, are in this subclass.

The third subclass is formed by decision making problems, in which majority of parameters are of quantitative nature, but there is no objective model for their aggregation. Many problems from these subclass are being solved with the help of multiattribute utility or value theory (Keeney, 1980, 1992; Keeney & Raiffa, 1976), by group of ELECTRE methods (e.g. Roy, 1985; Yu, 1993; Bana e Costa, 1992),and other methods (see Corner and Kirkwood, 1991).

Unstructured problems are the problems with the majority of qualitative parameters with unknown quantitative dependencies. Earlier (see Simon & Newell, 1958) it was considered that those tasks corresponded with purely heuristic ways of their solution. This means that there are no any ordered logical procedures for task solution. We can see examples of such tasks in policy making and strategic planning in different fields of human activities, in personal decision making.

Policy and analysis (Dror, 1989) and decision analysis (von Winterfeldt & Edwards, 1986; Watson & Buede, 1987; Goodwin & Wright, 1991) usually deal with such problems. Let us note the common features:

1. factors in these problems are of pure qualitative, subjective nature, especially difficult for formalization and measurement (prestige of the organization, attractiveness of the trade, attitude towards reforms, etc.);
2. process of the task analysis is also subjective by nature: rules for consideration and comparison of the main qualitative factors are mainly defined by the decision maker.
3. the decision maker prefers to use common for him (her) way of the problem description, process of its analysis and explanation of the obtained result;
4. a decision maker is the key element of the problem. This must be recognized, and attention thereof must be paid to the capabilities and limitations of human information processing system, to the results of investigations on human errors and heuristics (Kahneman et al., 1982).

Listed peculiarities of the unstructured problems show that the basic source of information, however, permitting the evaluation of a decision alternative, is a human being, i.e. decision maker. Hence, information elicitation must attend to the specifics and constraints of the human information processing system. It follows from this that the scientific criteria of decision method construction must be psychological criteria of "decision maker-method" interaction arrangement within a given decision method. This allows to define the following requirements for the decision making methods (Larichev, 1987, 1992):

(1) psychologically valid measurement of factors which are important for the decision;

(2) psychologically valid way of the eliciting information in the construction of a decision rule;

(3) possibility of checking the decision maker`s consistency;

(4) possibility for getting explanations.

Let us take a closer look at these requirements.

Measurements

Two first requirements are connected with the measurements. We call them primary ones (measurements of factors or criteria), and secondary ones (measurements of the decision maker's preferences).

In decision techniques , one may distinguish three groups of information processing operations: operations with criteria as items (for example, ordering the criteria by importance); operations with one alternative assessments by many criteria, (for example, to compare two estimates on two criteria of one alternative); operations with alternatives as items (for example, selecting the best alternative from several ones).

It was proposed (Larichev et al., 1987) to classify information processing operations into three groups:

- complex if psychological research indicates that in performing such operations the decision maker displays many inconsistencies, makes use of simplified strategies (e.g. drops a number of criteria);

- admissible if psychological research indicates that the decision maker is capable of performing them with small inconsistencies and using complex strategies (e.g., using combinations of criteria estimates);

- uncertain if not enough psychological research on these operations has been conducted. However one may be able to make a preliminary conclusion on admissibility or complexity of the operation.

The analysis demonstrates (Larichev et al., 1987) that there are not many admissible operations and all of them (but one) are of qualitative nature.

Therefore, while constructing psychologically valid decision methods it is necessary to use qualitative measurements (e.g., ranking of criteria by importance; comparison of verbal descriptions of scale values, and so on).

The results of descriptive research in decision making shows that it is important to use the appropriate language in carrying out measurements. Decision making methods should use pieces of natural for the decision maker language as a means for the problem description, for formulation of criteria and their scales, and for elicitation of preferences of a decision maker. Only admissible operations of information elicitation are to be used in psychologically valid decision aids.

Consistency test

One of the inherent characteristics of human behavior is error. In transmitting and processing information, people make errors. They make less and even considerably less errors when using valid information elicitation procedures described above, but all the same they do make errors. Hence, information obtained from a person must be subject to verification, rather than be used uncontrollably.

The decision making methods are to provide the possibility to check the information given by the decision maker for consistency and to eliminate contradictions in case they occur.

Generation of explanations

From a behavioral point of view, one of the requirements of the application of any method is its explainability. In making a crucial decision, the decision maker would like to know why alternative A turned out better than B, and both of them are better than C. This decision maker's requirement is quite reasonable. The stage of information elicitation from the decision maker (measurements) and the stage of final results presentation are separated by a stage of information transformation. Understandably, the decision maker wants to be sure that it is precisely his (or her) own preferences without any distortions that are behind the assessment of alternatives. In order to meet this requirement, the decision method must be "transparent": it must be conducive to finding a unambiguous correspondence between the information elicited from the decision maker and the final evaluations of alternatives. Only then there appears an opportunity to obtain explanations by the decision maker.

All requirements described above have been used as criteria for development of the decision making methods for unstructured problems. A set of such methods have been developed (Larichev, 1987; Larichev & Moshkovich, 1994a; 1994b, Berkeley et al., 1991). The description of one of them is given below.

METHOD 'ZAPROS' FOR UNSTRUCTURED PROBLEMS

Task formulation

To illustrate our approach, we shall describe here a method (and a system) ZAPROS, which is built upon the stated above principles, and show its applicability through an example.

ZAPROS is a multicriteria selection aid, based on qualitative judgments. It allows to construct a quasi-order on a rather large set of multiattribute alternatives using operation comparison of mulriattribute alternatives differing in values by two criteria. Let us consider the following problem.

There exists a Fund, which is meant to support financially good research projects. To organize the effective work of the Fund, the manager decided to elaborate a special form for submitting of the proposals, as well as a special form for referees, reviewing proposals. To be able to make quick decisions on projects after their reviewing, and to guarantee that projects are supported in accordance with the policy of the Fund, the manager formulated four main characteristics (important to him/her), and decided to construct a decision rule, which it would be possible to use for selection of an appropriate subset of the best proposals for financial support at each period of time. The elaborated set of characteristics with possible values on their scales are presented in Table 1.

Table 1. Attributes and possible values for evaluation of R&D projects

Attributes	Possible values on their scale
1. Originality	1. Absolutely new idea and/or approach 2. There are new elements in the proposal 3. Further development of previous ideas 4. Accumulation of additional data for previous research
2. Prospects	1. High probability of success 2. Success is rather probable 3. There is some possibility of success 4. Success is hardly probable
3. Qualification	1. Qualification of the proposer is high 2. Qualification of the proposer is normal 3. Qualification of the proposer is unknown 4. Qualification of the proposer is low
4.Level of the work	1. The proposed work is of high level 2. The proposed work is of middle level 3. The proposed work is of low level

The manager (the decision maker in this task) would like to formulate the decision rule, incorporating these attributes. Then, experts, while reviewing the proposals, can evaluate each proposal upon this set of attributes, and these evaluations will be taken into account in the decision. As the set of proposals is not known in advance, the idea is to construct some decision rule, which can be applied to any set of proposals.

The main idea of the method ZAPROS (Larichev & Moshkovich, 1991, 1994b) is based on the concept of a joint ordinal scale built according to the decision maker's preferences. The joint ordinal scale (JOS) means that all possible values upon all attributes (see Table 1 for example) are ranked-ordered for the decision maker upon his (or her) preferences. This ordinal scale may be effectively used (as it will be shown below) for pairwise comparison of alternatives.

Usually it is assumed that values upon each attribute scale are rank-ordered for the decision maker. In table 1 you can see that for each attribute scale values are rank-ordered from the most to the least preferable one. This means that the first value is more preferable that the second, etc.

To construct this JOS it is necessary to make 'ordinal trade-offs' for each pair of attributes and for each pair of possible values. Let introduce the following nominations for main elements of our task:

1. $K = \{1,2,...,Q\}$ - a set of attributes;

2. $\forall q \in K$, n_q- number of possible values on the scale X_q of the q-th attribute;

3. $\forall q \in K$, $X_q = 1,2,...,n_q$- where $\forall j,k \in X_q$: j is preferred to k if and only if $j < k$;

4. $Y = X_1 \times X_2 \times ... \times X_Q = \{ y=(i_1,i_2,...,i_Q) \mid \forall q \in K, i_q \in X_q \}$;

5. $|Y| = n_1 \cdot n_1 \cdot ... \cdot n_Q = N$;

To carry out trade-offs in an ordinal form we need to ask a decision maker questions of such kind: "What do you prefer: to have the best level upon attribute q and the second (in the rank order) level upon attribute q+1, or the best level upon attribute q+1 and the second (in the rank order) level upon attribute q?". For our example we have to ask: "What do you prefer in the proposal: to have an original idea with only probable success, or only some new elements with high probability of success?"

It is clear, that the analogous question when levels are changed from the best to the worst attribute level, correspond to the routine questions in the classical procedure for determination of attribute weights (Edwards & Winterfeldt, 1986), but does not require quantitative estimation of the preference.

These questions may be considered easy enough, but formulations are a bit too complicated. The same information may be obtained much easier in the form of comparison of two hypothetical alternatives, which differ in levels by only two attributes. This type of judgment was found valid (Larichev et al., 1987).

For our example we'll ask the decision maker to compare the following alternatives and with three variants of the possible answer (see Figure 1). As can be seen, values upon all attributes but the two under consideration, are at the same (best) level.

We can construct a set of hypothetical alternatives L, including vectors from Y for such comparisons. Their peculiarity is that they have the best attribute levels upon all attributes but one. Number of such alternatives is not large: $M = \sum_{q=1}^{Q}(n_q - 1) + 1$.

The set L may be defined formally as:

$L = \{ (1, 1, \ldots, 1, i_q, 1, \ldots, 1) \mid q \in K \text{ and } i_q \in X_q \}$.

For our example the set L will consist of the following 12 vectors: (1111), (2111), (3111), (4111), (1211), (1311), (1411), (1121), (1131), (1141), (1112), (1113), where vector (1211) means a hypothetical R&D proposal, which is characterized by the following values: absolutely original idea, success is rather probable, qualification of the proposer is high, and the proposed work is of high level.

The task is to fill in matrix of the size MxM with results of pairwise comparisons of alternatives from L.

```
================================================================
You are to compare the following alternatives:

ALTERNATIVE A (2111)                 ALTERNATIVE B (1211)

There are new elements in the proposal.  ←  Absolutely new idea and/or approach.
High probability of success.              →  Success is rather probable.
Qualification of the proposer is high.    =  Qualification of the proposer is high.
The proposed work is of high level.       =  The proposed work is of high level.

Possible answers:          1. A is preferred to B
                           2. A and B are indifferent
                           3. B is preferred to A.
================================================================
```

Figure 1. Visualization of alternatives for comparison

Construction of the joint ordinal scale

There are three basic sources for filling in the matrix. First, dominance relationships are imposed (due to ordinal scales). Next, decision makers can be to select their preferred choice from a pair of alternatives varying only on two attributes (in a discussed earlier trade off manner). Third, the system can infer other preferences through assumption of transitivity.

The dominance relation P^O is based on ordinality of scales and is defined as follows:

$P^O = \{ (y',y'') \in Y \times Y \mid \forall q \in K \ i'_q \leq i''_q$ and $\exists r \subset K$ such that $i'_r < i''_r \}$,

where $y' = (i'_1, i'_2, ..., i'_Q)$, $y'' = (i''_1, i''_2, ..., i''_Q)$.

Then all questions, necessary to compare all vectors from the list L, are asked (an example of such a question is presented in Figure 1). The results of pairwise comparisons by a DM may be presented in a form of binary relations as follows:

1) P_{DM} is a set of pairs $(y', y'') \in L \times L$, if according to a DM's opinion y is more preferable than y'' , or if $(y', y'') \in P^O$.

2) I_{DM} is a set of pairs $(y', y'') \in L \times L$, if according to a DM's opinion y and y are indifferent.

If we require the transitivity of relations P_{DM} and I_{DM} then according to Mirkin (1974) the relation $R_1 = P_{DM} \cup I_{DM}$ is a linear quasi-order on the set L.

Therefore, we need to construct a transitive binary relation R_1 on L.

In any interview with a DM there is a possibility of errors in his (her) responses. Therefore, a special procedure for detection and elimination of contradictions in the DM's responses is needed.

In the problem under consideration the possible contradictions in DM's responses may be determined as violations of transitivity of relations P_{DM} and I_{DM} (and in general as violations of transitivity of R_1).

The main idea of the proposed approach to the intransitivity detection and elimination (Moshkovich, 1988) is as follows. The transitivity of preferences assumes that if:

1) $(y', y'') \in P_{DM}$, then $\forall y''' \in L$ and $(y'', y''') \in P_{DM} \Rightarrow (y', y''') \in P_{DM}$;

2) $(y', y'') \in I_{DM}$, then $\forall y''' \in L$ and $(y'', y''') \in I_{DM} \Rightarrow (y', y''') \in I_{DM}$;

3) $(y', y'') \in P_{DM}$, then $\forall y''' \in L$ and $(y'', y''') \in I_{DM} \Rightarrow (y', y''') \in P_{DM}$;

4) $(y', y'') \in I_{DM}$, then $\forall y''' \in L$ and $(y'', y''') \in P_{DM} \Rightarrow (y', y''') \in P_{DM}$.

Therefore, after each comparison of vectors from L made by a DM, this information may be extended on the basis of transitivity (transitive closure of the binary relation defined on the set L is being built).

After that, the DM is presented with the next pair of vectors from L, for which the relation has not been defined. When the DM's response is obtained, the transitive closure is developed, and the procedure is maintained up to the moment of establishing relations for all pairs from L. It is proved that such a procedure does not lead to intransitivity of the relation being built (see Larichev and Moshkovich, 1994b).

To test responses of a decision maker, we suggest to present the DM with additional pairs of vectors for comparison on the basis of the following principle: the relation between each pair of vectors from L is to be defined directly (by a DM's response) or indirectly (by transitive closure) no less that two times. This requirement means that if a DM by two of his (her) responses (may be indirectly) has equally defined the relation between vectors from L, then this relation is considered to be proven. If the relation between two vectors from L has been defined only once, and only upon transitive closure, then this pair is presented additionally to a DM for comparison.

If the DM's response does not conflict with the previously obtained information, then the judgment is considered to be correct. If there is some difference, the triple of vectors for which a pairwise comparison contradicts the transitivity of the relation, built on L, is found out: that is of vectors y', y'', y''' ∈ L such that one of the following statements is fulfilled:

1) $(y', y'') \in P_{DM}$; $(y'', y''') \in I_{DM}$; $(y', y''') \in I_{DM}$;

2) $(y', y'') \in P_{DM}$; $(y'', y''') \in P_{DM}$; $(y', y''') \in I_{DM}$;

3) $(y', y'') \in P_{DM}$; $(y'', y''') \in P_{DM}$; $(y', y''') \in P_{DM}$;

4) $(y', y'') \in P_{DM}$; $(y'', y''') \in I_{DM}$; $(y', y''') \in P_{DM}$.

Such triple may always be detected, because after each of DM's responses we have built transitive closure of the obtained relation. In this case the DM is asked to reconsider the situation and to change one (or more) of his (or her) previous responses to eliminate intransitivity (example of such situation is presented in Figure 2).

After the corrected responses are obtained, they are incorporated into the information on the DM's preferences with introducing the necessary changes (for more details see Larichev & Moshkovich, 1994b).

The assumed transitivity of preferences and rank orderings of attribute levels make it possible to construct an effective procedure of pairwise comparisons, which essentially reduces the number of required comparisons from Mx(M-1)/2. For our example, the following 13 responses from a DM were enough to fill in the matrix of

ALTERNATIVE A
Absolutely new idea and/or approach
Qualification of the proposer is high
The proposed work is of low level

ALTERNATIVE B
There are new elements in the proposal
Qualification of the proposer is high
The proposed work is of high level

ALTERNATIVE C
Absolutely new idea and/or approach
Qualification of the proposer is unknown
The proposed work is of high level

/ alternatives have the best values upon other attributes/

Earlier you said that alternative A was preferred to B, alternative B was preferred to C. That's why A was preferred to C. Now you say that alternative C is preferred to B.

What comparison would you like to change ?

Possible answers : AB, BC, AC.

Figure 2. Visualization of a DM's contradictory responses

pairwise comparisons of vectors from L:

1. (2111) is preferred over (1211)

2. (2111) is preferred over (1121)

3. (1112) is preferred over (2111)

4. (1113) is preferred over (1211)
5. (3111) is preferred over (1211)
6. (1121) is preferred over (3111)
7. (3111) is preferred over (1131)
8. (1211) is preferred over (4111)
9. (4111) is preferred over (1311)
10. (1131) is preferred over (4111)
11. (4111) is preferred over (1141)
12. (1131) is preferred over (1211)
13. (1141) is preferred over (1311)

Then additional 3 pairwise comparisons were made to check the required information:

1. (1211) is preferred over (1141)
2. (1141) is preferred ovcr (1411)
3. (1121) is preferred over (1113)

This third answer contradicted to the previous ones (see figure 2), which in terms of vectors meant: (1113) was preferred over (2111); (2111) was preferred over (1121), thus (1113) was preferred over (1121). Now you said that (1121) was preferred over (1113).

The DM decided to change the relationship between (1113) and (2111). It was decided that (2111) was preferred over (1113). The system recalculated the matrix of pairwise comparisons and asked no more questions.

In experiments, carried out by Larichev et al. (1993) for the task with 5 attributes, four of which had 3 levels and one - 4 levels (that lead to M=11 and Mx(M-1)/2=55), the number of questions varied around 18-20, with maximum of 25.

Let us now analyze, how we are able to use the obtained judgments. As a result of an interview with a DM, the relation $R_1 = P_{DM} \cup I_{DM}$ of a linear quasi-order on the set L is built (see this rank-order for our example in Figure 3). This rank-order is called the joint ordinal scale (JOS).

We can formulate the following rule for comparison of a pair of vectors from Y:

<u>RULE 1</u>. Vector $y' = (i'_1 , i'_2 , ..., i'_Q) \in Y$ is not less preferable for the DM than vector $y'' = (i''_1 , i''_2 , ..., i''_Q) \in Y$, if for A attribute $q \in K$ E attribute $t(q) \in K$ such that:

$$(1 , 1, ..., 1, i'_q, 1, ..., 1) R_1 (1, 1, ..., 1,i''_{t(q)} , 1, ..., 1).$$

The correctness of such a rule in case of preferential independence of all pairs of attributes is proven in Larichev & Moshkovich (1991, 1994b).

Vectors from the set L differ from the best possible alternative in only one component. Therefore, we can consider the place, obtained by the vector in this ranking to be the place of this unique component in the JOS. That is why the rule for comparison of vectors from Y may be reformulated as follows:

<u>Reformulation of RULE 1</u>. Vector $y' \in Y$ is not less preferable than vector $y'' \in Y$, if for each component of the vector y there exists a component of the vector y with not more preferable value upon joint ordinal scale (binary relation R_1).

□ RANK	JOINT ORDINAL SCALE (ordered values)	Vector
1	Absolutely new idea and/or approach	1111
1	High probability of success	1111
1	Qualification of the proposer is high	1111
1	The proposed work is of high level	1111
2	The proposed work is of middle level	1112
3	There are new elements in the proposal	2111
4	The proposed work is of low level	1113
5	Qualification of the proposer is normal	1121
6	Further development of previous ideas	3111
7	Qualification of the proposer is unknown	1131
8	Success is rather probable	1211
9	Accumulation of additional data for previous research	4111
10	Qualification of the proposer is low	1141
11	There is some possibility of success	1311
12	Success is hardly probable	1411

Figure 3. Joint ordinal scale (ranking of vectors from list L)

The procedure of comparison of two vectors from Y may be fulfilled in the following way. Let us mark $r(i_q)$ the rank, which the vector $(1, 1, \ldots, 1, i'_q, 1, \ldots, 1)$ obtained in the joint ordinal scale. Then each vector $y' \in Y$ may be presented by vector of ranks $r' = (r(i'_1), r(i'_2), \ldots, r(i'_Q))$. We may rearrange ranks in a vector in a non-descending manner: the first $r'_1 = \min(r(i'_q))$, where q=1,2,...,Q; the second rank - the smallest from the rest and so on). As a result we have $r' = (r'_1, r'_2, \ldots, r'_Q)$ for y', and $r'' = (r''_1, r''_2, \ldots, r''_Q)$ for y", and y' is not less preferable than y", if $r'_q < r''_q$, q=1,2,...,Q.

This procedure guarantees the requirement of the RULE 1 (considering both formulations).

Let assume that in our example we have 12 proposals, estimated by experts, as shown in table 2 (the second column).

Table 2. Data on 12 alternative R&D proposals

Proposals	Attributes				Ranks in JOS				Rearranged ranks			
	Orig	Pros	Qual	Lev	Orig	Pros	Qual	Lev				
01	1	2	4	4	1	8	10	4	1	4	8	10
02	3	3	1	1	6	11	1	1	1	1	6	11
03	2	1	2	2	3	1	5	2	1	2	3	5
04	2	1	3	2	3	1	7	2	1	2	3	7
05	3	3	2	1	6	11	5	1	1	5	6	11
06	3	3	4	3	6	11	10	4	4	6	10	11
07	4	1	2	2	1	8	10	4	1	4	8	10
08	2	1	1	3	3	1	1	4	1	1	3	4
09	3	4	4	3	6	12	10	4	4	6	10	12
10	3	4	2	3	6	12	5	4	4	5	6	12
11	3	2	3	3	6	8	7	4	4	6	7	8
12	3	4	4	3	6	12	10	4	4	6	10	12

We change each component of a vector by the corresponding rank in the joint ordinal scale (figure 3). The result is presented in the third column of table 2. After that we rewrite these vectors with values in the non-descending order (see column 4 of the table 2). Now we are able to compare them. The result of comparison is presented in the matrix a) in figure 4.

This data may be used for rank-ordering of proposals. But let us recall, that this rule is correct when all pairs of attributes are mutually preferentially independent. Therefore, we need to discuss this problem in a more detail.

```
         3 1 2 2 1 3 2 3 3 3 3 3          3 1 2 2 1 3 2 3 3 3 3 3
         4 1 2 3 2 4 2 1 4 2 3 4          4 1 2 3 2 4 2 1 4 2 3 4
         2 3 1 1 3 3 1 1 4 4 2 4          2 3 1 1 3 3 1 1 4 4 2 4
         1 3 2 2 3 3 4 2 3 3 3 3          1 3 2 2 3 3 4 2 3 3 3 3

1.  1243 2 3 0 0 3 1 0 0 1 3 3 1          2 3 0 0 3 3 1 0 1 1 3 1
2.  3311   2 3 3 1 1 3 0 1 1 3 1            2 3 3 1 1 3 3 1 1 3 1
3.  2122     2 1 1 1 1 0 1 1 1 1              2 1 3 1 1 0 1 1 1 1
4.  2132       2 1 1 1 0 1 1 1 1                2 3 1 1 0 1 1 1 1
5.  3321         2 1 9 9 1 1 3 1                  2 1 3 3 1 1 3 1
6.  3343           2 0 0 1 3 0 1                    2 0 0 1 1 0 1
7.  4122             2 0 1 1 3 1                      2 0 1 1 3 1
8.  2113               2 1 1 1 1                        2 1 1 1 1
9.  3443                 2 0 0 2                          2 0 0 2
10. 3423                   2 3 1                            2 0 1
11. 3233                     2 1                              2 1
12. 3443                       2                                2

         a)                                     b)
```

1 - alternative in the row is preferred to alternative in the column; 2 - alternatives in the row and the coluumn are equally preferable; 0 - alternative in the row is less preferable than alternative in the column; 3 alternatives in the row and the column are in comparable.

Figure 4. Matrices of pairwise comparisons of proposals:a) upon JOS for list L; b) upon JOS2 for list L_2 .

Preferential independence of attributes

First let us remind the notion of preferential independence (Keeney, 1974).

Definition 1. Attributes s and t of the set K are preferentially independent from the other attributes of this set, if preference between vectors with equal values upon all attributes but s and t, does not depend on the values of the equal components.

In practical problems we must check if this axiom is not violated in DM's preferences. The problem of checking this axiom (as well as checking many other axioms of multiattribute utility theory) has no simple solution. In reality, the necessity to use this axiom results from the desire to construct an effective decision rule on the basis of relatively small amount of rather simple information about DM's preferences (the effectiveness of the decision rule means its possibility to guarantee rather high level of compatibility for real alternatives). On the other hand, the full-scale check of DM's preferences implies the need for a DM to carry out a large number of pairwise comparisons. So, the point is to make not a full-scale but sufficient check of DM's preferences to satisfy the axiom's conditions. The following approach is proposed.

Let us form list L_2 of vectors from the set Y with all the least preferable components, but one. The list L_2 for our example will consist of the following 12 vectors: (1443), (2443), (3443), (4143), (4243), (4343), (4413), (4423), (4433), (4441), (4442), (4443). Analogously to the relation R_1 we are able to build the relation of linear quasi-order R_2 on the set L_2. This relation may be used to make some verification of the preferential independence of attributes.

For our example the 14 responses of a DM allowed to fill in the matrix for vectors from L_2. In the end we received JOS2, presented in figure 5.

First let us show, how we can use the received comparisons for checking the preference independence of attributes. List L_2 contains vectors with all values but one, equal to the worst ones, and with one value at the best level. So, there is a possibility to compare relations between pairs of vectors from the two lists of the following type:

L_1: $(1, 1, \ldots, 1, n_s, 1, \ldots, 1)$ and $(1, 1, \ldots, 1, n_t, 1, \ldots, 1)$

L_2: $(n_1, n_2, \ldots, n_{s-1}, 1, n_{s+1}, \ldots, n_Q)$ and $(n_1, n_2, \ldots, n_{t-1}, 1, n_{t+1}, \ldots, n_Q)$

Both pairs of vectors differ only in components upon attributes s and t. So, pairs differ from one another only in values of equal components. Therefore, if attributes s and t are preferentially independent, the preference in the pairs from the lists L and L_2 has to be

RANK	JOINT ORDINAL SCALE (ordered values)	Vector
1	High probability of success	4143
2	Success is rather probable	4243
3	There are new elements in the proposal	2443
4	There is some probability of success	4343
5	Qualification of the proposer is high	4413
6	Qualification of the proposer is normal	4423
7	The proposed work is of high level	4441
8	Absolutely new idea and/or approach	1443
9	Further development of previous ideas	3443
10	Qualification of the proposer is unknown	4433
11	The proposed work is of middle level	4442
12	Accumulation of additional data for previous research	4443
12	Qualification of the proposer is low	4443
12	Success is hardly probable	4443
12	The proposed work is of low level	4443

Figure 5. Joint ordinal scale (ranking of vectors from list L_2)

the same. This is a possibility to carry out some justification of the axiom on the basis of the JOS and JOS2. Let us analyze this possibility with the help of our example: comparison of vectors (4143) and (4413) from the list L_2 has to be the same as for vectors (1141) and (1411) from the list L, if attributes 2 and 3 (Prospects and Qualification) are preferentially independent of attributes 1 and 4 (Originality and Level) according to the definition 3.

As can be seen, in reality the decision maker is to make the same trade-off between values upon attributes 2 and 3 in both cases.

Let us emphasize that though such justification is a very limited one, the violation of this condition rather clearly proves the violation of independence and the necessity of additional analysis of the situation (see later), as all these relations have been thoroughly checked during comparisons for lists L and L_2. Additionally, let us note that the selected lists of vectors differ to a very large extent (in the quality of presented vectors), so the correspondence of the obtained results may be considered to be stable and for all intermediate situations.

It is easy to prove that the introduced above Rule 1 for comparison of vectors from Y may be modified for implementation on the basis of the relation R_2.

RULE 2. If each pair of attributes from K (Q>3) does not depend preferentially on other attributes, then vector $y' = (i'_1, i'_2, \dots, i'_Q) \in Y$ is not less preferable for a DM, than vector $y'' = (i''_1, i''_2, \dots, i''_Q) \in Y$, if for $\forall s \in K \; \exists \; t(s) \in K$ that:

$(n_1, n_2, \dots, n_{s-1}, i'_s, n_{s+1}, \dots, n_Q) \, R_2 \, (n_1, n_2, \dots, n_{t-1}, i''_t, n_{t+1}, \dots, n_Q)$ And if attributes $s,q \in K$ are such that $s \neq q$, then $t(s) \neq t(q)$.

Therefore, while comparing real alternatives it is possible to use joint ordinal scales, built for both lists of vectors. If the results are the same, then this is an indirect confirmation of preferential independence of attributes and justification of rules being used (this means that we carry out additional check of attribute independence just while comparing real alternatives).

If according to RULE 1 alternatives can be compared and according to RULE 2 we are not able to compare them, then this is not a contradictory situation. We have just enlarged compatibility of alternatives on the basis of additional information from a DM about comparison of vectors from L_2 .

If the results of comparison contradict each other, then this is connected with violation of attribute independence. The alternatives are to be considered incomparable. So, even if there is some evidence about the dependency of some pair of attributes, we are able to use the built rules for comparison those pairs of alternatives, uninfluenced by this dependency. We are able to estimate the number of alternatives' pairs which it will be possible to compare additionally if we analyze the dependency thoroughly. Analysis and elicitation of dependent attributes and also procedures for reformulation of the initial task in this case (Larichev & Moshkovich, 1991) are rather labor-consuming. So, a DM is able to evaluate if he (or she) wants to spend rather large amount of time and effort, knowing the maximum of additional information about comparison of alternatives which it is possible to obtain as a result.

It is clear that differences in comparison of vectors from Y on the basis of the JOS built for two lists of vectors, are caused only by the information about DM's preferences presented in relations R_1 and R_2. Let us prove the following statement.

Statement. Let the comparison of vectors $(y', y'') \in Y$ on the basis of the relation R_1 and the relation R_2 be different. Then there always exist attributes s and t for which:

$(1, 1, ..., 1, i'_s, 1, ..., 1) R_1 (1, 1, ..., 1, i''_t, 1, ..., 1)$;

but $(n_1, n_2, ..., n_{t-1}, i''_t, n_{t+1}, ..., n_Q) R_2 (n_1, n_2, ..., n_{s-1}, i'_s, n_{s+1}, ..., n_Q)$.

(Proof of the statement is given in APPENDIX 1)

This statement allows us to carry out the check of the preferential independence of attributes on the basis of R_1 and R_2 due to the following corollary.

Definition 2. Let us call pairs of vectors from L and L_2 analogous ones, if they have the same components, different from the best or the worst ones correspondingly. This means that $(b_s, b_t) \in L$ is analogous to $(b'_s, b'_t) \in L_2$, if:

$b_s = (1, 1, ..., 1, i'_s, 1, ..., 1)$;

$b_t = (1, 1, ..., 1, i''_t, 1, ..., 1)$;

$b'_s = (n_1, n_2, ..., n_{s-1}, i'_s, n_{s+1}, ..., n_Q)$;

$b'_t = (n_1, n_2, ..., n_{t-1}, i'_t, n_{t+1}, ..., n_Q)$.

Corollary. If comparisons between all analogous pairs of vectors from lists L and L_2 are the same then it is impossible to detect violations of preferential independence of attributes on the basis of the obtained information. (The proof is evident, if to consider the above marked possibilities to detect violations of the axiom about preferential independence of attributes).

This gives the opportunity to find out pairs of dependent attributes by analyzing corresponding comparisons for two lists of vectors. To use additional information we compared our proposals (from table 2) on the basis of the JOS2. The resulting matrix is presented at figure 4b. In analyzing the two matrices, we found out that we are able to add only three additional comparisons to the matrix at figure 4a (they are underlined in the matrix at figure 4b). Now we have full information, checked for errors and violations of the axiom, and are able to use this information for rank-ordering of our alternatives.

In case of dependency, we advise to reconstruct the attributes (merge and/or additionally subdivide them) to avoid this dependency. And then to repeat the procedure (for more details, see Larichev & Moshkovich, 1991).

Rank-ordering of alternatives

Analyzing the matrix in figure 4a, we can see rather large number of incomparable alternatives. This leads us to the question of how to rank-order alternatives on the basis of this matrix. There are several approaches to this problem. We advise to provide possibilities to rank order alternatives upon basic principles, adopted by specialists in data analysis. The most popular four of them are the following:

1 - sequential selection of non-dominated alternatives;

2 - sequential selection of non-dominating alternatives;

3 - sequential selection of alternatives, which dominate maximum of other alternatives;

4 - sequential selection of alternatives, which dominate minimum of other alternatives.

The idea of the first principle is to select the first group of non-dominated alternatives (containing the best one). After that these alternatives are excluded from the set, and once more a set of non-dominated alternatives out of this subset is selected. These alternatives are considered to be of the second rank. And so on. This procedure can be said to work in the form up-down.

The second principle works in a reverse manner. We first select alternatives, which do not dominate any other alternatives. This subset is considered to be the least preferable one. After that these alternatives are excluded from the initial set of alternatives and the following subset of not-dominating alternatives is selected.

The third principle is based on the sequential selection of alternatives, which dominate the largest number of other alternatives. And the fourth principle is a reverse to the third one and sequentially selects alternatives, which are dominated by the largest number of other alternatives, considering the first selected group to be the least preferable one.

These four principles of ranking of alternatives on the basis of partial information about their pairwise comparisons represent the basic ideas in this field and may be easily applied to the same matrix of pairwise comparisons.

To analyze the results of our example we used all four principles. In figure 6 the graphs, representing these rank orderings are presented.

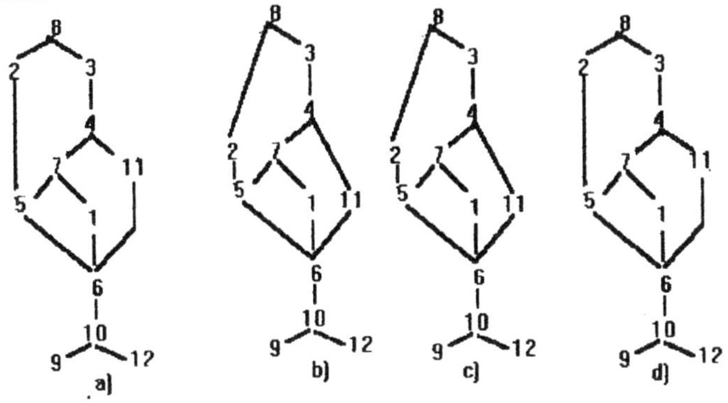

Figure 6. Rank-orderings of alternatives upon different principles: a) principle of sequential selection of non-dominated alternatives; b) principle of sequential selection of non-dominating alternatives; c) principle of sequential selection of alternatives, which dominate maximum of other alternatives; d)principle of sequential selection of alternatives, which dominate minimum of other alternatives.

As we can see, though we are not able to form a full ranking of alternatives, we can form rather stable partial ordering, which may be effectively used in decision making. There is a stable leader in this group (the proposal 8), then we can elaborate a group of "better alternatives", consisting of alternatives 2, 3 and 4; the next group is consisted of alternatives 1, 5, 7 and 11. Though we have some additional information about comparison of some alternatives within these groups, this is not crucial for the decision making in this task, as the aim is to support proposals of good quality (not just the best one). The decision maker was satisfied with the result, and agreed that this information is enough for the decision making.

Explanations

There is an easy possibility to get explanations for comparison of any two alternatives in the built ranking according to the presented above rule. If these alternatives are incomparable on the basis of this rule, then the resulting relation between them is explained with the help of additional alternatives.

Let alternative y' has smaller rank than alternative y" in the final ranking. At the same time on the basis of the JOS alternatives y' and y" are incomparable. Then if the ranking was done according to the principle of sequential selection of non-dominated or non-dominating alternatives, the alternative y'" is searched for, which is incomparable with alternative y' but is more preferable than alternative y". If the ordering is according to the number of dominated (dominating) alternatives, then the alternative y'" is searched for, which is dominated by y' but is not dominated by y".

Examples of possible messages for explanation of the results of comparison of real alternatives are given in figure 7.

CONCLUSION

The necessary element of multiattribute decision methods is the process of information elicitation from decision makers and experts. Only this information is able to eliminate the uncertainty which is connected with the presence of multiple criteria, to

a)

Alternative 7 (4122) IS MORE PREFERABLE THAN Alternative 6 (3343)

because as a result of the interview it is stated that:

value 4 upon attribute 1 (alt. 7) IS MORE PREFERABLE THAN
value 3 upon attribute 2 (alt. 6);

value 1 upon attribute 2 (alt. 7) IS MORE PREFERABLE THAN
value 3 upon attribute 2 (alt. 6);

valuc 2 upon attribute 3 (alt.7) IS MORE PREFERABLE THAN
value 4 upon attribute 3 (alt.6);

value 2 upon attribute 4 (alt.7) IS MORE PREFERABLE THAN
value 3 upon attribute 4 (alt.6);

b)

Alternative 2 (3311) IS MORE PREFERABLE THAN Alternative 1 (1243)

because:

they are incomparable on the basis of JOS

but the least preferable alternative 8 dominating alternative 2,

HAS SMALLER RANK THAN

the least preferable alternative 4 dominating alternative 1.

Figure 7. Possible explanations of comparisons of alternatives: a) comparison upon JOS; b) comparison upon the principle of ranking.

elaborate the necessary compromise and find good decisions. There is evidence that it is more preferable to use human judgments in a qualitative form: it is more natural for people, and provides more reliable information. We can see that there are possibilities to use qualitative information in a logical (theoretically correct) way to compare and evaluate multiattribute alternatives.

Quantitative measurement of qualitative notions, which dominate in unstructured problems, may lead to an incorrect result, which it is difficult to detect. In this case we achieve only the impression that we have the decision, as we substitute a decision maker by some (smaller) decisions of a consultant (or the author of the method). In general, unessential differences in numerical expression of values and weights, as a rule, may not confirm decision makers in validity of the applied methodology.

It is necessary to understand that in many cases we are able to obtain a task solution in practice without the resort to numerical scaling. We have shown an a example of a real decision task, which does not need to rank order all alternatives, or to choose the best one. In this case the proposed procedure ZAPROS, based on ordinal trade offs may lead to a satisfactory, easily explainable and trustable solution. Such forms of human judgment allow to carry out logical analysis of the received information, detect possible inconsistencies and to overcome them through an additional analysis.

APPENDIX 1

Proof of the statement.

As it is true that y' is preferred to y" $\in$ Y on the basis of relation R_1, then according to RULE 2 for $\forall s \in K \; \exists \; t(s)$ e K such that:

$(1, 1, ..., i'_s , 1, ..., 1) R_1 (1, 1, ..., i''_t , 1, ..., 1)$.

As the relation R_2 is a connected one, then:

(1) $(n_1, n_2, ..., n_{s-1}, i'_s, n_{s+1}, ..., n_Q) R_2 (n_1, n_2, ..., n_{t(s)-1}, i''_t, n_{t(s)+1}, ...,n_Q)$;

or:

$$(2)\ (n_1, n_2, \ldots, n_{t(s)-1}, i''_t, n_{t(s)+1}, \ldots, n_Q)\ R_2\ (n_1, n_2, \ldots, n_{s-1}, i'_s, n_{s+1}, \ldots, n_Q).$$

If (2) is true, then the statement is proved. So, let it be that for each $s \in K$ the condition (1) is fulfilled. Then according to RULE 2 y' is preferred to y" on the basis of the relation R_2 and this contradicts to the initial conditions of the statement.

REFERENCES

Bana e Costa, C.A., "Absolute and relative evaluation problematics. The concept of neutral level and the MCDA Robot Technique", in Cerny, D.Gluckaufova and D. Loula (eds), *Proceedings of the International Workshop on Multicriteria Decision Making, Methods - Algorithms - Applications Liblice, March 18-22, 1991 (Prague 1992).*

Brown R.V. (1989) Toward a prescriptiive science and technology of decision aiding. - *Annals of operations research*, 19, 467-483.

Corner J.L. and Kirkwood, C.W. (1991) Decision analysis applications in the operations research literature, 1970-1989. - *Operations Research*, 39, 206-219.

Dror Y. (1989) Public Policy making Reexamined. - Transaction Publishers: New Brunswick (USA) and Oxford (UK).

Goodwin P. and Wright G., (1991) Decision analysis for management judgments.- Y.:Wiley.

Kahneman D., Slovic P., Tversky A. (eds) (1982), Judgment under uncertainty: heuristics and biases. - Cambridge: Univ. Press.

Keeney R.L. (1974), Multiplicative utility functions. - *Operation Research*, 22, 22-34.

Keeney R.L. (1980), Siting energy facilities. - N.Y.: Academic Press.

Keeney R., (1992) Value-focused thinking. - Harvard Univ. Press.

Keeney R., Raiffa H. (1976), Decisions with multiple objectives: preference and value tradeoffs.- New York: Wiley.

Larichev O.I. (1987) Objective models and Subjective Decisions. - Moscow: Nauka, 87 (in Russian).

Larichev O.I. (1992), Cognitive validity in design of decision-aiding techniques. - *Journal of Multi-Criteria Decision Analysis*, 1, 127-138.

Larichev O.I., Furems E.M. (1987) Multicriteria Packing Problems. In: Y.Sawaragi, K.Inoue, H.Nakayama (Eds) *Towards Interactive and Intelligent Decision Support Systems*, Springer-Verlag.

Larichev O.I., Kozucharov A. (1979) Multiple criteria assignment problem: combining the collective criterion with individual preferences. - *Mathematical Science Humain*, 68, 63-77.

Larichev O.I., Moshkovich H.M. (1988) Limits to decision making ability in direct multiattribute alternative evaluation. - *OBHDP*, 42, 217-233.

Larichev O., Moshkovich H. (1990) Decision Support System CLASS for R&D Planning, - *Proceedings of First International Conference on Expert Planning Systems, Brighton*, Conf. Public. N 322, The Institue of Electrical Engineers, 227-232.

Larichev O.I., Moshkovich H.M. (1991), *ZAPROS: a method and a system for ordering of multiattribute alternatives on the base of decision-maker's preferences*. - Preprint, Moscow: VNIISI press, 1991 (in English).

Larichev O.I., Moshkovich H.M. (1992) Decision Support System ORCLASS, *Proceedings of the Tenth International Conference on Multiple Criteria Decision Making*, Taiwan, 1, 341- 350.

Larichev O.I., Moshkovich H.M. (1994a) An Approach to Ordinal Classification Problems. - *Int.Trans.Oper.Res.*, v.1,no.3,pp.375-385.

Larichev O.I., Moshkovich H.M. (1994b) ZAPROS-LM a method and system for ordering multiattribute alternatives. - *European Journal of Operational Research* (in press).

Larichev O.I., Moshkovich H.M., Mechitov A.I., Olson D.L. (1993), Qualitative approaches to rank-ordering multiattribute alternatives. - *Journal of Multi-Criteria Decision Analysis*, 2, 5-26.

Larichev O.I., Polyakov O.A., Nikiforov A.D. (1987), Multicriterion linear programming problems (Analytical Survey) - *Journal of economical psychology*, 8, 389-407.

Larichev O.I., Zuev Yu.A., Gnedenko L.S. (1974), Method for classification of applied R&D projects. - In: S.V.Emelyanov (ed.) *Perspective apllied R&D planning*. - Moscow: Nauka press, 28-57 (in Russian).

Mirkin B.G. (1974), Problems of group choice. - Moscow: Nauka press (in Russian).

Moskovich H.M. (1988) Interactive system ZAPROS (for ordering of multiattribute alternatives on the basis of decision maker's preferences). - In: Larichev O.I., Emelyanov S.V. (eds), *Man-machine procedures for decision making*. - Moscow: VNIISI press, 11, 13-21 (in Russian).

von Neumann J.& Morgenstern O. (1947) Theory of games and economic behavior. - Princeton University Press, Princeton.

Quade E.S. (1984), Analysis of Military Decisions. - Chicago: McNally.

Roy B. (1985) Methodologe multicritere d'aide a la decision, Economica.

Simon H., Newell A. (1958), Heuristic Problem Solving: the next Advance in Operastions Research. - *Operations Research*, 6.

Jan. Steuer R. (1986) Multiple Criteria Optimization: Theory, Computation and Application. N.Y., Wiley.

von Winterfeldt D. and Edwards W., (1986), Decision Analysis and Behavioral Research. - Cambridge University Press.

Yu, W., (1992), ELECTRE TRI, Aspects methodologiques et manuel d'utilisation, in:Document du LAMSADE, 74 (Paris: Universite de Paris-Dauphine).

MULTICRITERIA ANALYSIS IN PROBLEM SOLVING AND STRUCTURAL COMPLEXITY

Victor Korotkich

The Computing Center of the Russian Academy of Sciences, Vavilov str. 40, 117967, Moscow, Russia

e-mail: korot@sms.ccas.msk.su

Introduction

Many complex problems that arise in knowledge processing, control theory and optimization have the following generic form.

A problem solver takes in information in the form of a data stream, i.e., receives serially a sequence of data $u_1 u_2 ... u_n, n \geq 2$. At each k-th, $k = 2, 3, ..., n-1$ step the problem solver is provided with information $(v_1, u_1, v_2, u_2, ..., v_{k-1}, u_{k-1})$ and selects a strategy v_k from a given class of permissible strategies. Problem solving has a top-level goal that can be satisfied by satisfying subgoals at each step. Selection of a strategy at the k-th step is defined by the k-th step subgoal satisfaction. At the end of the k-th step the problem solver, with the k-th strategy v_k selected, receives the data u_k which provide a possibility of evaluating a condition of the k-th $k = 1, 2, ..., n$ subgoal satisfaction.

Multicriteria analysis arises naturally in problem solving. The purpose of the analysis is to find regularities in data stream, elicit relevant knowledge and compress it into a model to improve problem solving [1]. This logical process requires a range of criteria by which new facts and beliefs are derived from known facts and beliefs to make prediction of the new strategy and modification of the model as experience accumulates [2],[3].

The purpose of this paper is to focus attention on the role of a structural complexity as a tool for the criteria definition and study of optimal problem solving. The particular interest lies on a study of a connection between optimal problem solving and a symmetry.

Implementing this program we reduce all of problem solving to a specific form, i.e., to a model of binary sequence prediction with the number of correctly determined symbols as an objective function. In the model a binary sequence is considered as an arbitrary, individual entity without any assumptions on the existence of an underlying statistics.

It turns out that all algorithms in the context of the model are equivalent, i.e., they have the same performance. But the model reveals an algorithm which possesses the most symmetry. By exploring this "symmetric" algorithm we try to shed light on certain aspects of optimality.

Namely, we introduce a notion of structural optimality and prove that the algorithm is structural optimal. This result uncovers the role of the complexity in problem solving as it specifies the criteria used by the optimal algorithm.

The structure of the paper is as follows. In §1 to set the stage for the study, we review a structural complexity and results involved. In §2 and §3 we give

P.M. Pardalos et al. (eds.), Advances in Multicriteria Analysis, 81–90.

the model and algorithm descriptions respectively. In §4 structural optimality of the algorithm is shown.

§1. Integer code series and binary sequence structural complexity. An integer code series is the basic tool to the concepts definition used herein.

Let $\Delta_n[t_m, t_{m+n}]$ denote the partition of an interval $[t_m, t_{m+n}] \in \Re^1$ with points $t_k, k = m, m+1, ..., m+n$ such that $t_m = m\varepsilon, |m| = 0, 1, \dots, \varepsilon > 0$, $n = (t_{m+n} - t_m)/\varepsilon$ and $t_{k+1} - t_k = \varepsilon, k = m, m+1, ..., m+n-1$. Let $F(\Delta_n[t_m, t_{m+n}])$ be the set of all piecewise constant functions$f : [t_m, t_{m+n}] \to \Re^1$ such that each f is constant on $(t_k, t_{k+1}], k = m, m+1, ..., m+n-1$. The sequence $s_1 s_2 ... s_n$, where s_k is f value on $(t_{m+k-1}, t_{m+k}], k = 1, 2, ..., n$ and $f(t_m) = s_1$ is called a code of f and is denoted by $c(f)$. The integer code series reads

Theorem 1[4]. Let $f \in F(\Delta_n[t_m, t_{m+n}])$, $c(f) = s_1 s_2 ... s_n$. Then the r - th $r = 1, 2, ...$ integral value of f at t_{m+k}, $k = 1, 2, ..., n$ satisfies the formula

$$f^{[r]}(t_{m+k}) = \sum_{i=0}^{r-1} \alpha_{rmi}((m+k)^i s_1 + ... + (m+1)^i s_k)\varepsilon^r + \sum_{i=1}^{r} \beta_{rki} f^{[i]}(t_m)\varepsilon^{r-i}, \quad (1)$$

where $\alpha_{rmi} = C_r^i((-1)^{r-i-1}(m+1)^{r-i} + (-1)^{r-i} m^{r-i})/r!$, $i = 0, 1, ..., r-1$, C_r^i - the binomial coefficient and $\beta_{rki} = k^{r-i}/(r-i)!$, $i = 1, 2, ..., r$.

The main concept which provides the basis of our construction is a binary sequence structural complexity[5].

Let the map ρ_m take each sequence $s = s_1 s_2 ... s_n, s_k \in \Re^1, k = 1, 2, ..., n$ to$f \in F(\Delta_n[t_m, t_{m+n}])$ such that $c(f) = s_1 s_2 ... s_n$ and $f^{[i]}(t_m) = 0, i = 1, 2, \dots$. Let $B_n = \{\sigma : \sigma = \sigma_1 \sigma_2 ... \sigma_n, \sigma_k = \{-1, 1\}, k = 1, ..., n\}$ be the set of all binary sequences of length n. Length $|\sigma|$ of σ is the number of symbols (bits) in σ and $\sigma(k) = \sigma_1 \sigma_2 ... \sigma_k, k = 1, 2, ..., n$. Consider (1) for $f = \rho_m(\sigma), \sigma \in B_n$ and $t_{m+n}, \varepsilon = 1$. Then (1) takes the form

$$f^{[r]}(t_{m+n}) = \sum_{i=0}^{r-1} \alpha_{rmi}((m+n)^i \sigma_1 + (m+n-1)^i \sigma_2 + ... + (m+1)^i \sigma_n). \quad (2)$$

The powers $(m+n)^i, (m+n-1)^i, ..., (m+1)^i, i = 0, 1, ..., r-1$ are included in (2) with + or - signs dependent on the associated component of $c(f)$. Thus, for a binary sequence $\sigma \in B_n$ there is a decomposition of $\{m+1, m+2, ..., m+n\}$ into two subsets. By $x_k \in \{m+1, m+2, ..., m+n\}$ denote a number such that its associated code component is the k-th (counting from left) component equals 1, by y_i - equals -1. Let $X_m(\sigma) = (x_1, x_2, ..., x_p)$, $Y_m(\sigma) = (y_1, y_2, ..., y_q)$, ($X_m(\sigma) = \emptyset$ if $p = 0$, $Y_m(\sigma) = \emptyset$ if $q = 0$), $0 \le p, q \le n$, $p + q = n$,

$$\aleph_i(\sigma, m) = (x_1^i + x_2^i ... + x_p^i - y_1^i - y_2^i - ... - y_q^i) = \sum_{j=1}^{p} x_j^i - \sum_{j=1}^{q} y_j^i, i = 0, 1, \dots \quad .$$

Then (2) takes the form

$$f^{[r]}(t_{m+n}) = \sum_{i=0}^{r-1} \alpha_{rmi}(x_1^i + x_2^i + ... + x_p^i - y_1^i - y_2^i - ... - y_q^i) = \sum_{i=0}^{r-1} \alpha_{rmi}\aleph_i(\sigma, m).$$

Since $f^{[r]}(t_{m+n}), r = 1, 2, ...$ are invariant with m, it can be thought of as $m = 0$ ($\rho_0 = \rho, X_0(\sigma) = X(\sigma), Y_0(\sigma) = Y(\sigma), \aleph_i(\sigma, 0) = \aleph_i(\sigma), i = 0, 1, ...$).

A structural representation of $\sigma \in B_n$ is the sequence $\mathbf{S}(\sigma) = \{S_i(\sigma) = f^{[i]}(t_n), i = 1, 2, ...\}$ of $f = \rho(\sigma)$ successive integrals at the point t_n.

Definition 1. Let $\sigma \in U \subseteq B_n$. A structural complexity $\mathbf{C}(\sigma, U)(\mathbf{C}(\sigma, B_n) = \mathbf{C}(\sigma))$ of σ with respect to U is the number of $\mathbf{S}(\sigma)$ first successive components that is necessary and sufficient for σ unique determination in U.

Consider for $X(\sigma) = (x_1, x_2, ..., x_p), Y(\sigma) = (y_1, y_2, ..., y_q)$ a system of Diophantine equations $D_k(\sigma), k = 1, 2, ...$

$\mathbf{x}_1^0 + \mathbf{x}_2^0 + ... + \mathbf{x}_p^0 - \mathbf{y}_1^0 - \mathbf{y}_2^0 - ... - \mathbf{y}_q^0 = \aleph_0(\sigma)$

$\mathbf{x}_1^1 + \mathbf{x}_2^1 + ... + \mathbf{x}_p^1 - \mathbf{y}_1^1 - \mathbf{y}_2^1 - ... - \mathbf{y}_q^1 = \aleph_1(\sigma)$

. . .

$\mathbf{x}_1^{k-1} + \mathbf{x}_2^{k-1} + ... + \mathbf{x}_p^{k-1} - \mathbf{y}_1^{k-1} - \mathbf{y}_2^{k-1} - ... - \mathbf{y}_q^{k-1} = \aleph_{k-1}(\sigma)$

in integers such that $1 \leq \mathbf{x}_i \leq n, i = 1, 2, ..., p$, $1 \leq \mathbf{y}_i \leq n, i = 1, 2, ..., q, \mathbf{x}_i \neq \mathbf{x}_j, i \neq j,\ i, j = 1, 2, ..., p, \mathbf{y}_i \neq \mathbf{y}_j, i \neq j,\ i, j = 1, 2, ..., q,\ \mathbf{x}_i \neq \mathbf{y}_j, i = 1, 2, ..., p, j = 1, 2, ..., q$.

Theorem 2[6]. The structural complexity $\mathbf{C}(\sigma)$ of a binary sequence $\sigma \in B_n$ equals the minimal number k such that $D_k(\sigma)$ possesses a unique solution.

By using results from [7],[8], the structural complexity $\mathbf{C}(\sigma)$ of σ can be also expressed by the genus $\mathbf{g}$ of the corresponding smooth projective curve, i.e., it satisfies the inequality $\mathbf{C}(\sigma) \leq \mathbf{g} + 1$ and it is conjectured, that $\mathbf{C}(\sigma) = \mathbf{g} + 1$ [6].

A sequence $\sigma' \in B_{k'}$ is called a sequence $\sigma \in B_n, n = 0 mod(k')$ generator and is denoted by $\sigma' = gen(\sigma)$ if $\sigma = \underbrace{\sigma'...\sigma'}_{n/k'} = (\sigma')^{n/k'}$ and there exists no $\sigma'' \in B_{k''}$ such that $\sigma' = \underbrace{\sigma''...\sigma''}_{k'/k''} = (\sigma'')^{k'/k''}, k' = 0 mod(k'')$ for some integer $k'' > 1$.

Consider a set of binary sequences $\Omega_n^1 = \{\omega_k, k = 1, ..., N+1\} \subset B_n, n = 2^N, N = 1, 2, ...$. The first element $\omega_1 \in \Omega_n^1$ is $(1)^n$. Adjacent elements $\omega_k, \omega_{k+1} \in \Omega_n^1, k = 1, ..., N$ are produced by generators $\omega'_k = gen(\omega_k), \omega'_{k+1} = gen(\omega_{k+1})$ with relation $\omega'_{k+1} = \omega'_k \bar{\omega}'_k$ (in $\bar{\omega}'_k$ all 1 are replaced by -1 and -1 by 1).

The initial segment of length n of the Thue-Morse sequence τ (begining with 1) is defined as $\tau(n) = \omega_{N+1}(n)$. The Thue-Morse sequence (begining with -1) equals $\bar{\tau}$.

The sequences τ, $\bar{\tau}$ with respect to a structural symmetry [9] have remarkable properties.

Theorem 3 [4]. Let $X_m(\tau(n)) = (x_1, x_2, ..., x_p)$, $Y_m(\tau(n)) = (y_1, y_2, ..., y_q)$, $n = 2^N, N = 1, 2, ...$, $0 \leq p, q \leq n$, $p + q = n$. Then
$x_1^0 + x_2^0 + ... + x_p^0 = y_1^0 + y_2^0 + ... + y_q^0$
$x_1^1 + x_2^1 + ... + x_p^1 = y_1^1 + y_2^1 + ... + y_q^1$
. . . .
$x_1^{N-1} + x_2^{N-1} + ... + x_p^{N-1} = y_1^{N-1} + y_2^{N-1} + ... + y_q^{N-1}$.
The same statement is valid for $\bar{\tau}$.

For example, for $\tau(16)$, $m = 6$, rearranging for some elegance, we have
$7^0+10^0+12^0+13^0+16^0+17^0+19^0+22^0 = 8^0+9^0+11^0+14^0+15^0+18^0+20^0+21^0$
$7^1+10^1+12^1+13^1+16^1+17^1+19^1+22^1 = 8^1+9^1+11^1+14^1+15^1+18^1+20^1+21^1$
$7^2+10^2+12^2+13^2+16^2+17^2+19^2+22^2 = 8^2+9^2+11^2+14^2+15^2+18^2+20^2+21^2$
$7^3+10^3+12^3+13^3+16^3+17^3+19^3+22^3 = 8^3+9^3+11^3+14^3+15^3+18^3+20^3+21^3$

It is on the basis of these properties of the symmetry that we make the following Conjecture. First, we introduce the following definition.

Definition 2. Let $\sigma, \sigma' \in B_n, \sigma \neq \sigma'$. The number $k, k = 0, 1, ..., \mathbf{C}(\sigma) - 1$ such that $S_1(\sigma) = S_1(\sigma'), S_2(\sigma) = S_2(\sigma'), ..., S_k(\sigma) = S_k(\sigma'), S_{k+1}(\sigma) \neq S_{k+1}(\sigma')$ ($k = 0$, if $S_1(\sigma) \neq S_1(\sigma')$) is called a structural similarity of σ to σ' and is denoted by $\Delta\mathbf{S}(\sigma, \sigma')$.

Clearly $\Delta\mathbf{S}(\sigma, \sigma') = \Delta\mathbf{S}(\sigma', \sigma)$. According to this definition sequences have more structural similarity the longer their initial structural representation components agree.

Conjecture 1. Let $\sigma \in B_n$, $\sigma \neq \tau(n), \bar{\tau}(n)$, $n = 2, 3, ...$. Then

$$\Delta\mathbf{S}(\sigma, \bar{\sigma}) < \Delta\mathbf{S}(\tau(n), \bar{\tau}(n)) = log_2 n = \mathbf{C}(\tau(n)) - 1, \tag{3}$$

if $n = 2^N, N = 1, 2, ...$ and

$$\Delta\mathbf{S}(\sigma, \bar{\sigma}) \leq \Delta\mathbf{S}(\tau(n), \bar{\tau}(n)), \tag{4}$$

if $n \neq 2^N, N = 1, 2, ...$.

§2. The model of problem solving.

As a basis for the study of problem solving we reduce all of it to a specific form, i.e., to the case when $v_k, u_k \in \{-1, 1\}, k = 1, 2, ..., n$. It means that the problem solver at the k-th, $k = 1, 2, ..., n$ step makes correct prediction of the strategy (to satisfy the k-th subgoal) when $v_k = u_k$ and doesn't make correct prediction when $v_k = \bar{u}_k$. In the last case the strategy $\bar{v}_k$ is required to be selected to satisfy the k-th step subgoal.

Problem solving in this form is supposed to predict what the next binary symbol of an input sequence will be and make analysis on how the sequence is structured, based on the past succession of symbols.

A deterministic n-step sequential algorithm $A \in \boldsymbol{A}$ is a set $\{\chi_k, k = 1, 2, ..., n\}$ of functions each having the form $v_1 = \chi_1 = const$, $v_k = \chi_k$ $(v_1, u_1, v_2, u_2, ..., v_{k-1}, u_{k-1})$, $u_k, v_k \in \{-1, 1\}, k = 1, 2, ..., n$, where $\mathbf{A}$ is the set of all such algorithms. An algorithm A takes in information serially in the form of an input binary sequence $u = u_1 u_2 \dots u_n$, and at each k-th step determines, by using χ_k, the value of v_k which is the evaluation of the value of u_k, $k = 1, 2, \dots, n$. The value of v_k depends, in general, on the values of $v_1, u_1, v_2, u_2, \dots, v_{k-1}, u_{k-1}$ on the previous steps, and is determined by A before the value of u_k, $k = 1, 2, \dots, n$ becomes known. After n steps $u = u_1 u_2 \dots u_n$ gives rise to $v = A(u) = v_1 v_2 \dots v_n$ produced by A.

A priori knowledge of input sequences is provided through a model. In the model any input sequence u is considered as an arbitrary, individual entity without any assumptions on the existence of an underlying statistics.

Let $\mathbf{N}(u, A)$ be the number of steps at which A correctly determines the value of u symbol. This criterion is natural when dealing with algorithms whose goal is to correctly determine input symbols as many times as possible. Implicit in the criterion is the assumption that all correct determinations are equivalent.

Traditionally, two approaches are used to evaluate algorithms, worst case and average case performances. An algorithm A_* is said optimal, if it maximizes

$$\min_{u \in B_n} \mathbf{N}(u, A_*) = \max_{A \in \mathbf{A}} \min_{u \in B_n} \mathbf{N}(u, A)$$

and average optimal, if it maximizes

$$\sum_{u \in B_n} \mathbf{N}(u, A_*) = \max_{A \in \mathbf{A}} \sum_{u \in B_n} \mathbf{N}(u, A),$$

where the sum is taken over all $u \in B_n$.

The following evident lemmas are needed for the sequel.

Lemma 1. Let $A = \{\chi_k, k = 1, 2, \dots, n\} \in \mathbf{A}$ be an arbitrary algorithm. Then

$$\min_{u \in B_n} \mathbf{N}(u, A) = 0.$$

Proof. We prove by induction on $k, k = 1, 2, \dots, n$ that there exists $u = u_1 u_2 \dots u_n \in B_n$ such that $v = A(u)$, and $v_k = -u_k$, $k = 1, 2, \dots, n$. Clearly, this means that $\mathbf{N}(u, A) = 0$, which is what the lemma claims.

For $n = 1$, by definition $v_1 = \chi_1 = const$. Let $u_1 = -\chi_1 = -v_1$. Since $v_1 = -u_1$, the result follows.

Assume there exists $u(k) = u_1 u_2 \dots u_k, k = 1, 2, \dots, n-1$ such that $v_i = \chi_i(v_1, u_1, v_2, u_2, \dots, v_{i-1}, u_{i-1}) = -u_i, i = 1, 2, \dots, k$. Then at the $k+1$-th step of the algorithm $v_{k+1} = \chi_{k+1}(v_1, u_1, v_2, u_2, \dots, v_k, u_k)$ and we can choose u_{k+1} such that $u_{k+1} = -v_{k+1}$. This completes the proof.

Lemma 2. Let $A = \{\chi_k, k = 1, 2, \dots, n\} \in \mathbf{A}$ be an arbitrary algorithm. Then

$$\sum_{u \in B_n} \mathbf{N}(u, A) = n 2^{n-1}.$$

Proof. We first show that A correctly determines 2^{n-1} values of $u_k, k = 1, 2, \dots, n$ symbol when summing over all input sequences. Since, for $k = 1$, there are 2^{n-1} input sequences $u = u_1 u_2 \dots u_n$ such that $u_1 = \chi_1, \chi_1 \in \{-1, 1\}$, then it follows that 2^{n-1} values of u_1 symbol are correctly determined by the algorithm.

Clearly, at the k-th, $k = 2, 3, ..., n-1$ step there are 2^{k-1} different $u(k-1) = u_1u_2...u_{k-1}$ segments of all input sequences. Then, for some $u(k-1)$, we have $v(k-1) = v_1v_2...v_{k-1}$, $v_i = \chi_i(v_1, u_1, v_2, u_2, ..., v_{i-1}, u_{i-1}), i = 1, 2, ..., k$. Since there are 2^{n-k} input sequences $u' = u'_1u'_2...u'_n \in B_n$ such that $u'_i = u_i, i = 1, 2, ..., k-1$ and $u_k = v_k$, then 2^{n-k} values of u'_k symbol are correctly determined. Summing over all 2^{k-1} segments, one obtains 2^{n-1} symbols of u_k symbol are correctly determined at the k-th step. Finally, summing over all n steps, we have that the number of correctly determined symbols for all $u \in B_n$ equals $n2^{n-1}$. This completes the proof.

¿From Lemmas 1 and 2 it follows that all algorithms are equivalent, i.e., they have the same performance. The model assumptions and criterion are insufficient to give rise to the optimal algorithms.

There are a lot of extensions of the model, with optimal algorithms development being the crucial issue [10]-[13].

What really matters with the model is that it reveals an algorithm which possesses the most structural symmetry. By exploring this symmetric algorithm we try to shed light on certain aspects of optimality.

Some intuitive reasoning can be provided for this approach. Consider additional assumptions are putting (in some sense) parametrically into the model or the criteriion is slowly modifying. Then, imagine optimal algorithms are "appearing" and "moving" under such changes in the algorithms set. We conjecture, roughly speaking, that the motion starts in the symmetric algorithm and is very much connected with it.

Now we shall give the following definition.

Let $A \in \mathbf{A}, u \in B_n, v = A(u)$. For the pair (u, v) we define inductively a pair of binary sequences $(u \ominus v, v \ominus u)$ as follows. If there exists an index i_1 such that $u_j = v_j, 1 \leq j \leq i_1 - 1$, $u_{i_1} = -v_{i_1}$, if $1 < i_1 \leq n$, and $u_1 = -v_1$, if $i_1 = 1$, then $(u \ominus v)_1 = u_{i_1}$. Else, let $u \ominus v$ be the empty sequence.

Assume the first $k, 1 \leq k < n$ symbols of $u \ominus v$ are defined, i.e., $(u \ominus v)(k) = (u \ominus v)_1(u \ominus v)_2...(u \ominus v)_k$. If there exists an index i_{k+1} such that $u_j = v_j, i_k+1 \leq j \leq i_{k+1} - 1$, $u_{i_{k+1}} = -v_{i_{k+1}}$, if $i_k + 1 < i_{k+1} \leq n$, and $u_{i_{k+1}} = -v_{i_{k+1}}$, if $i_{k+1} = i_k + 1$, then $(u \ominus v)_{k+1} = u_{i_{k+1}}$. Else, let $u \ominus v = (u \ominus v)(k)$. Set $u \ominus v = -(v \ominus u)$.

The binary sequence $u \ominus v$ is formed consecutively from u symbols which were not correctly determined by A and the goal of its definition is to focus attention on these symbols.The algorithm A instead of $u \ominus v$ produces $v \ominus u$ and we are looking to measure how similar it is to $u \ominus v$.

The basic idea to address this problem is to use the similarity $\mathbf{\Delta S}(u \ominus v, v \ominus u)$ of $u \ominus v$ to $v \ominus u$ to define a criterion. It characterizes A ability to discover structural regularities in $u \ominus v$. A justification for this definition is the following. Despite the fact that no symbol in $u \ominus v$ was correctly determined by A, at the same time $\mathbf{\Delta S}(u \ominus v, v \ominus u), 0 \leq \mathbf{\Delta S}(u \ominus v, v \ominus u) \leq \mathbf{C}(u \ominus v) - 1$ of its structural(global) properties were determined by A. Note that, $\mathbf{C}(u \ominus v)$ of these properties uniquely determine $u \ominus v$.

An algorithm A_* is called (average) structural optimal, if it maximizes
$\sum \Delta\mathbf{S}(u \ominus v, v \ominus u) = max \sum \Delta\mathbf{S}(u \ominus v, v \ominus u) =$
$u \in B_n$ $A \in \mathbf{A} u \in B_n$
$v = A_*(u)$ $v = A(u)$
This performance measures the number of all $u \ominus v$ structural properties found by A over all $u \in B_n$ and A_* maximizes this quantity.

§3. Description of the symmetric algorithm.

We suggest an algorithm A_c which has the following description. On the first step the algorithm A_c set up $v_1 = -1$. On the k-th, $k = 2, ..., n$ step of the algorithm we have

1. if $u(k-1) \ominus v(k-1) = \emptyset$, then $v_k = v_{k-1}$,
2. if $u(k-1) \ominus v(k-1) \neq \emptyset$,

then by construction $u(k-1) \ominus v(k-1) \not\sim v(k-1) \ominus u(k-1)$. This means that there exists $\kappa, 1 \leq \kappa \leq \mathbf{C}(u(k-1) \ominus v(k-1))$ such that $S_1(u(k-1) \ominus v(k-1)) = S_1(v(k-1) \ominus u(k-1)), S_2(u(k-1) \ominus v(k-1)) = S_2(v(k-1) \ominus u(k-1)), ..., S_{\kappa-1}(u(k-1) \ominus v(k-1)) = S_{\kappa-1}(v(k-1) \ominus u(k-1)), S_\kappa(u(k-1) \ominus v(k-1)) \neq S_\kappa(v(k-1) \ominus u(k-1))$.

If $S_\kappa(u(k-1) \ominus v(k-1)) - S_\kappa(v(k-1) \ominus u(k-1)) > 0$, then $v_k = 1$, if $S_\kappa(u(k-1) \ominus v(k-1)) - S_\kappa(v(k-1) \ominus u(k-1)) < 0$, then $v_k = -1$.

Therefore, the structural complexity enters the realm of the analysis in problem solving in the following role. The algorithm A_c at each step uses different number κ of criteria, which increases as $\mathbf{C}(u(k-1) \ominus v(k-1))$ becomes greater.

Further we will need the following result about the algorithm A_c.

Theorem 4 [6]. Let $u \in B_n$ and at the k-th,$k = 2, 3, ...n$ step $|u(k) \ominus A_c(u(k))| = i, i = 1, 2, ..., k$. Then

$$u(k) \ominus A_c(u(k)) = \tau(i), \tag{5}$$

and if there exists an algorithm A that satisfies (5), then $A = A_c$.

§4. Structural optimality of the symmetric algorithm.

When the model is considered it is possible to obtain the following result.

Let $\mathbf{A}^- = \{A \in \mathbf{A}, \chi_1 = -1\}$.

Theorem 5. The algorithm A_c is the unique structural optimal algorithm in $\mathbf{A}^-$.

Proof. One step in the proof is to note that any algorithm A given the number of $u \ominus v, u \in B_n, v = A(u)$ such that $|u \ominus v| = i$ equals $C_n^i, i = 0, 1, ..., n$. We prove this statement by induction on $n = 1, 2, ...$.

For $n = 1$, since by A_c definition $v_1 = -1$, then we get one sequence $u \ominus v$ such that $|u \ominus v| = 1$, when $u_1 = -v_1 = 1$, and one sequence $u \ominus v$ such that $|u \ominus v| = 0$, when $u_1 = v_1 = -1$. Since $C_1^0 = 1$ and $C_1^1 = 1$, then it proves the statement for $n = 1$.

Induction hypothesis. Assume the truth of the statement for $n = k$. This means the number of $u \ominus v$ such that $|u \ominus v| = i$ equals $C_k^i, i = 0, 1..., k$. This assumption given we prove the statement for $n = k + 1$.

In fact, at the k-th step, let $|u(k) \ominus v(k)| = i$. It is clear that at the $k+1$ -th step there are two sequences one of which has length i and the other length $i+1$. Indeed, if case $v_{k+1} = 1$ occurs, then for $u_{k+1} = 1$ and $u_{k+1} = -1$ we have respectively $|u(k+1) \ominus v(k+1))| = |u(k) \ominus v(k)| = i$, and $|u(k+1) \ominus v(k+1)| = |u(k) \ominus v(k)| + 1 = i + 1$.

Analogously, if case $v_{k+1} = -1$ occurs, then we get $|u(k+1) \ominus v(k+1)| = |u(k) \ominus v(k)| + 1 = i+1$ for $u_{k+1} = 1$, and $|u(k+1) \ominus v(k+1)| = |(u(k) \ominus v(k)| = i$ for $u_{k+1} = 1$.

Thus, every sequence $u(k) \ominus v(k)$ such that $|u(k) \ominus v(k)| = i$ gives rise to two sequences of length i and $i+1$. In its turn, at the $(k+1)$-th step all sequences $u(k+1) \ominus v(k+1)$ such that $|u(k+1) \ominus v(k+1)| = i$ are resulted from all sequences $u(k) \ominus v(k)$ such that $|u(k) \ominus v(k)| = i$, $|u(k) \ominus v(k)| = i - 1$.

By the induction hypothesis, the number of $u(k) \ominus v(k)$ such that $|u(k) \ominus v(k)| = i, i = 0, 1, ..., k$ equals C_k^i. Therefore, the number of $u(k+1) \ominus v(k+1)$ such that $|u(k+1) \ominus v(k+1)| = i$ equals $C_k^i + C_k^{i-1} = C_{k+1}^i$ for $i = 1, 2, ..., k$, and C_k^i for $i = 0, k+1$. Since $C_k^0 = C_{k+1}^0$ and $C_k^k = C_{k+1}^{k+1}$, then the induction step is completed and the statement is proved.

For the performance $\mathbf{P}(A)$ of any algorithm $A, v = A(u)$ we have

$$\mathbf{P}(A) = \sum_{u \in B_n} \Delta\mathbf{S}(u \ominus v, v \ominus u). \tag{6}$$

Represent (6) as

$$\mathbf{P}(A) = \sum_{i=0}^{n} \sum_{|u \ominus v| = i} \Delta\mathbf{S}(u \ominus v, v \ominus u), \tag{7}$$

where all sequences are grouped together in accordance with length $i, i = 0, 1, ..., n$ and then are summed over all possible length values.

Let $\mathbf{n} = \{2, 3, ..., n\}$ and $\tilde{\mathbf{n}} = \{2, 4, ..., 2^{[log_2 n]}\}$, where $[log_2 n]$ is the integer part of $log_2 n$. By using the Conjecture conditions (3) and (4) respectively, we can rewrite (7) in the form

$$\mathbf{P}(A) = \sum_{i \in \tilde{\mathbf{n}}} \sum_{|u \ominus v| = i} \Delta\mathbf{S}(u \ominus v, v \ominus u) + \sum_{i \in \mathbf{n} \setminus \tilde{\mathbf{n}}} \sum_{|u \ominus v| = i} \Delta\mathbf{S}(u \ominus v, v \ominus u) <$$

$$\sum_{i \in \tilde{\mathbf{n}}} \sum_{|u \ominus v| = i} \Delta\mathbf{S}(\tau(i), \bar{\tau}(i)) + \sum_{i \in \mathbf{n} \setminus \tilde{\mathbf{n}}} \sum_{|u \ominus v| = i} \Delta\mathbf{S}(\tau(i), \bar{\tau}(i)). \tag{8}$$

It follows from the statement proved above that for any algorithm A

$$\sum_{|u \ominus v| = i} \Delta\mathbf{S}(\tau(i), \bar{\tau}(i)) = C_n^i \Delta\mathbf{S}(\tau(i), \bar{\tau}(i)). \tag{9}$$

By using (9) we obtain for (8)

$$\mathbf{P}(A) < \sum_{i \in \tilde{\mathbf{n}}} C_n^i \Delta\mathbf{S}(\tau(i), \bar{\tau}(i)) + \sum_{i \in \mathbf{n} \setminus \tilde{\mathbf{n}}} C_n^i \Delta\mathbf{S}(\tau(i), \bar{\tau}(i)) =$$

$$= \sum_{i=0}^{n} C_n^i \Delta\mathbf{S}(\tau(i), \bar{\tau}(i)) \tag{10}$$

By (5) in Theorem 4 and the statement, we can rewrite (10) as

$$\mathbf{P}(A) < \sum_{i=0}^{n} C_n^i \Delta\mathbf{S}(\tau(i), \bar{\tau}(i)) = \sum_{i=0}^{n} \sum_{|u \ominus A_c(u)|=i} \Delta\mathbf{S}(u \ominus A_c(u), A_c(u) \ominus u) =$$

$$= \sum_{u \in B_n} \Delta\mathbf{S}(u \ominus A_c(u), A_c(u) \ominus u) = \mathbf{P}(A_c).$$

This completes the proof of the theorem.

Acknowledgement - The research was supported by grant 93-012-450 Russian Scientific Fund.

REFERENCES

1. F. Hayes-Roth, N. Jacobstein, The State of Knowledge Based Systems. Communications of the ACM, March 1994, Volume 37, Number 3, 27-39.
2. A. Newell, H. Simon, Human Problem Solving. Prentice-Hall, Englewood Cliffs, N.J.,1972.
3. J. Pearl, Heuristics: Intelligent Search Strategies for Computer Problem Solving. Addison-Wesley, Reading, Mass.,1984.
4. V. Korotkich, An Approach to Nonlinear Systems Cycles Properties Description. Preprint, Institute for Automated Systems, Moscow, 1988(in russian).
5. V. Korotkich, Integer Code Series with Some Applications in Dynamical Systems and Complexity. Preprint, The Computing Center of the Russian Academy of Sciences, Moscow 1993.
6. V. Korotkich, Towards Symbolic Description of Complex Adaptive Systems Self-Organizational Processes. Notes of the Russian Academy of Sciences,Technical Cybernetics,N1,1994, pp.20-31.
7. A. Weil, Number of Solutions of Equations in Finite Fields. Bull. Amer. Math. Soc. 55, pp. 497-508(1949).
8. P. Deligne, La Conjecture de Weil. I,Publ.Math.IHES,43, pp. 273-307(1974).
9. V. Korotkich, Chaos Quantization to Relation Between Algorithm Optimality and Symmetry. Proceedings of the International Conference on Dynamical Systems and Chaos, Tokyo, May, 1994, Pergamon Press.
10. N. Merhav, Feder M, Universal Schemes for Sequential Decision from Individual Data Sequences. IEEE Transactions on Information Theory, Vol. 39, No. 4, July 1993, 1280-1292.
11. P. Algoet, Universal Schemes for Prediction, Gambling, and Portfolio Selection, Ann. Probab., Apr. 1992.

12. T. Cover, Universal Gambling Schemes and the Complexity Measures of Kolmogorov and Chaitin, Tech. Rep. 12, Dept. of Statist., Stanford Univ., 1974.
13. G. Goodwin, R. Payne, Dynamic System Identification: Experiment Design and Data Analysis, Vol.136, New York, Academic Press, 1977.

III. UTILITY ASSESSMENT

GENERAL OVERVIEW OF THE MACBETH APPROACH

Carlos A. Bana e Costa
Technical University of Lisbon
IST – Dep. Civil Eng. / CESUR
Av. Rovisco Pais, 1000 Lisbon, Portugal

Jean-Claude Vansnick
University of Mons-Hainaut
F.S.E.S., Place du Parc, 20
7000 Mons, Belgium

INTRODUCTION

MACBETH is an interactive approach for cardinal measurement of judgments about the degrees to which the elements of a finite set A possess a property P. The name *MACBETH*, ***Measuring Attractiveness by a Categorical Based Evaluation Technique***, comes from the fact that we conceived our approach with the aim of facilitating the measurement of (degrees of) attractiveness in decision processes. Nevertheless, MACBETH can also be applied to measure other properties in domains of knowledge others than Decision Sciences, such as in Psychophysics or in Social Sciences.

In this article we offer a general overview of MACBETH where we introduce some modifications improving the initial technical formulation of [Bana e Costa and Vansnick, 1993 and 1994a], although the basic conceptual ideas remain the same.

Objectives and questioning mode of MACBETH

Let A be a finite set of potential actions for which an actor D wants our aid in quantifying his judgments of attractiveness of the elements of A. In such a decision context, the MACBETH approach facilitates the construction of a numerical scale $v: A \rightarrow \Re: a \rightarrow v(a)$ such that:

- Not only, $v(a)$ numerically represents the attractiveness of action a for D in the (substantive) sense that:

(1) $\forall a,b \in A$, $v(a) > v(b)$ if and only if D judges a more attractive than b $(a\,P\,b)$

- but also, the positive difference $v(a) - v(b)$ numerically represents the difference of attractiveness between the actions a and b for D in the sense that:

(2) $\forall a,b,c,d \in A$ with $a\,P\,b$ and $c\,P\,d$, $[v(a)-v(b)]/[v(c)-v(d)]$ reflects the ratio (that D feels with greater or lesser precision) of the differences of attractiveness between a and b on the one hand and c and d on the other hand.

Obviously, such a scale v satisfies the conditions defining the notion of "measurable value function" (cf. [Dyer and Sarin, 1979]), for from **(2)** immediately follows that $\forall a,b,c,d \in A$ with $a\,P\,b$ and $c\,P\,d$, $v(a)-v(b) > v(c)-v(d)$ if and only if the difference of attractiveness between a and b is larger than the difference of attractiveness between c and d.

P.M. Pardalos et al. (eds.), Advances in Multicriteria Analysis, 93–100.

Several techniques have been suggested in the literature for the construction of such a value function, but the difficulty of the questions that D is asked to answer makes that notion hardly operational in practice. One of the key advantages of MACBETH is precisely that its questioning procedure is particularly intelligible and straightforward, for it involves only two actions in each question. For this purpose, MACBETH makes use of the *notion of difference of attractiveness* between two actions of A.

In fact, $\forall\ a, b \in A$ with $a\,Pb$, MACBETH's questioning mode consists of asking D to express an *absolute judgment* of difference of attractiveness by assigning the pair (a, b) to one of three semantic categories:

C_2 $\rightarrow$ ***weak*** *difference of attractiveness*
C_4 $\rightarrow$ ***strong*** *difference of attractiveness*
C_6 $\rightarrow$ ***extreme*** *difference of attractiveness*

or, in the case of hesitation, to one of three intermediate categories:

C_1 $\rightarrow$ ***very weak*** *difference of attractiveness (from nil to weak)*
C_3 $\rightarrow$ ***moderate*** *difference of attractiveness (from weak to strong)*
C_5 $\rightarrow$ ***very strong*** *difference of attractiveness (from strong to extreme).*

From a mathematical point of view, the categories C_1, C_2, C_3, C_4, C_5 and C_6 are asymmetric binary relations which constitute a partition of $P = \{(a,b) \in A \times A \mid D$ judges a more attractive than $b\}$.

The notion of absolute judgment has also been used by Saaty [1977, 1980] in the questioning mode of his Analytical Hierarchy Process (AHP). In spite of this common basic idea, there is significant fundamental differences between MACBETH and AHP. First, in MACBETH the absolute judgments concern *differences* (of attractiveness) whereas in Saaty's method they concern *ratios* (of priority, or of importance). Moreover, and this a key distinction between the technical procedures behind the two approaches, for determining a numerical scale on A, Saaty *a priori* associates a single arbitrarily fixed number to each of his categories of ratios, whereas in MACBETH:

1. we associate to each of our six categories C_1 to C_6 an <u>interval</u> of the real line, as shown in figure 1, and
2. the thresholds s_i (i = 1, 2, 3, 4, 5, 6) defining the intervals associated with our categories <u>are not</u> *a priori* fixed; on the contrary, they are determined simultaneously with the numerical scale v that we are looking for.

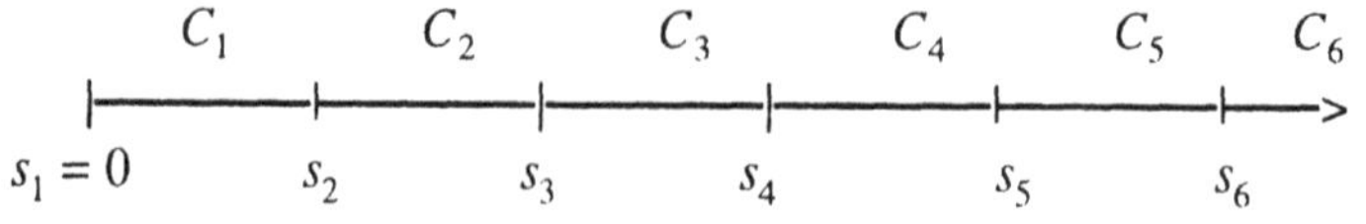

Figure 1 – Representation of MACBETH's categories in the real line

More accurately, once D has assigned each pair $(a, b) \in P$ to one of the six semantic categories of difference of attractiveness, MACBETH tries to determine simultaneously:

- an *application* v: $A \to \Re$ associating each element a of A with a real number $v(a)$ and
- six real numbers s_1, s_2, s_3, s_4, s_5 and s_6 such that

(3) $$0 = s_1 < s_2 < s_3 < s_4 < s_5 < s_6$$

(4) $\forall a, b \in A$:

$$\begin{cases} \forall k \in \{1, 2, 3, 4, 5\}: \quad s_k < v(a) - v(b) < s_{k+1} \text{ if and only if } (a, b) \in C_k \\ \text{and} \qquad s_6 < v(a) - v(b) \text{ if and only if } (a, b) \in C_6 . \end{cases}$$

Let us see how MACBETH interacts with D to reach this purpose, assuming, for reasons of clarity of exposition, that the matter here is only the <u>cardinal</u> component of the problem. That is to say that we will assume the following hypotheses:

Hypothesis H1 The binary relation P modelling the ordinal judgments of attractiveness of D about the elements of A, is asymmetric, negatively transitive, and connected (that is, P is a total strict order).

Hypothesis H2 $A = \{a_1, a_2, \ldots, a_{n-1}, a_n\}$ such that

$$\forall i \neq j \in \{1, 2, \ldots, n\}, a_i \, P \, a_j \text{ if and only if } i > j$$

that is, the complete ranking of the elements of A by order of decreasing attractiveness is $a_n \, P \, a_{n-1} \, P \ldots P \, a_2 \, P \, a_1$.

In these conditions, to collect D's absolute judgments of difference of attractiveness between the elements of A it is sufficient to fill in, row by row, the upper part of the $n \times n$ matrix

	a_n	a_{n-1}	.	.	.	a_2	a_1
a_n		$a_{n,n-1}$	.	.	.	$a_{n,2}$	$a_{n,1}$
a_{n-1}			.	.	.	$a_{n-1,2}$	$a_{n-1,1}$
.				.	.	.	.
.					.	.	.
.						.	.
a_2							$a_{2,1}$
a_1							

where, $\forall i > j \in \{1, 2, \ldots, n\}$ and $\forall k \in \{1, 2, 3, 4, 5, 6\}$:

$$a_{i,j} = k \text{ if and only if } (a_i, a_j) \in C_k.$$

Cardinal consistency

In theory, the search for an application $v: A \to \Re$ satisfying **(4)** is a particular case of the problem of the *constant threshold representation* of a *m*-tuple of binary relations, studied by Doignon [1987] who stated a necessary and sufficient condition (cyclone condition) for the existence of a solution.

Bana e Costa and Vansnick [1994b] prove that, when hypothesis H1 is satisfied (*P* is a total strict order), Doignon's condition is a necessary and sufficient one for the existence of an application $v: A \to \Re$ satisfying **(4)** and **(3)**. In practice, it is unfortunately very difficult to test the theoretical condition of Doignon. It is the reason why we developed another path for verifying the existence of a solution for our problem. It consists in solving the following linear program (called *Mc*1), the variables of which are $v(a)\ (a \in A)$, $s_1, s_2, s_3, s_4, s_5, s_6$ and c:

Min c

s.t.

r0) all variables ≥ 0

r1) $s_1 = 0$

r2) $v(a_1) = 0$ (see hypothesis H2: $\forall a \in A,\ aPa_1$)

r3) $\forall k \in \{2, 3, 4, 5, 6\}$: $s_k - s_{k-1} \geq 1000$

r4) $\forall k \in \{1, 2, 3, 4, 5, 6\},\ \forall (a,b) \in C_k : v(a) - v(b) \geq s_k + 1 - c$

r5) $\forall k \in \{1, 2, 3, 4, 5\},\ \forall (a,b) \in C_k : v(a) - v(b) \leq s_{k+1} - 1 + c$

Bana e Costa and Vansnick [1994b] prove that an application $v: A \to \Re$ satisfying **(3)** and **(4)** exists if and only if the optimal solution c_{min} of *Mc*1 is equal to 0. When $c_{min} \neq 0$ there exists some <u>cardinal inconsistency</u> in the absolute judgments of *D*, in the sense that the desired numerical representation is not possible. In this case, it is important to try to identify the judgments that are causing trouble in order to discuss with *D*. Two linear programs, called *Mc*3 and *Mc*4, have been developed for that purpose.

Some new non negative variables intervene in *Mc*3, together with $v(a)$ ($a \in A$), s_1, s_2, s_3, s_4, s_5 and s_6: $\alpha(a,b)$ and $\delta(a,b)$ $[(a,b) \in C_1 \cup C_2 \cup \ldots \cup C_6]$ and $\beta(a,b)$ and $\gamma(a,b)$ $[(a,b) \in C_1 \cup C_2 \cup \ldots \cup C_5]$. These variables are defined by:

r6) $\forall k \in \{1, 2, 3, 4, 5, 6\}$ and $\forall (a,b) \in C_k$: $v(a) - v(b) = s_k + 1 - \alpha(a,b) + \delta(a,b)$,

r7) $\forall k \in \{1, 2, 3, 4, 5\}$ and $\forall (a,b) \in C_k$: $v(a) - v(b) = s_{k+1} - 1 + \beta(a,b) - \gamma(a,b)$.

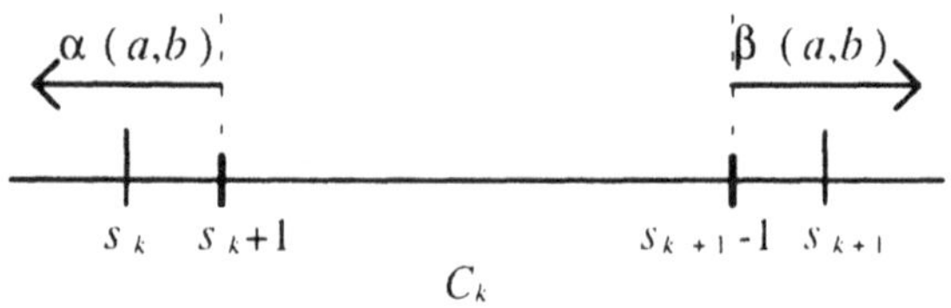

Figure 2: Variables $\alpha(a,b)$ and $\beta(a,b)$ for k < 6

The variables $\alpha(a,b)$ and $\beta(a,b)$ (see figure 2) are particularly interesting for they enable the immediate identification of the ordered pairs (a,b) that can be the cause of inconsistency. The objective function is the same in $Mc3$ and $Mc4$:

$$\text{Min}\left[\sum_{(a,b)\in C_1\cup C_2\cup\ldots\cup C_6} \alpha(a,b) \;+\; \sum_{(a,b)\in C_1\cup C_2\cup\ldots\cup C_5} \beta(a,b)\right]$$

and the constraints are:

- for $Mc3$: $r0$, $r1$, $r2$, $r3$, $r4$ and $r5$ (both with c replaced by c_{min}), $r6$ and $r7$;
- for $Mc4$: $r0$, $r1$, $r2$, $r3$, $r6$ and $r7$.

Note that the two programs only differ on the constraints $r4$ and $r5$ which are not included in $Mc4$. The aim of $Mc3$ is to detect all the possible sources of inconsistency; in particular, it can be useful to identify the cyclones related to the theoretical condition of Doignon (see Bana e Costa and Vansnick [1994b]). $Mc4$ suggests a restricted number of modifications of category to tentatively reach cardinal consistency. Let us mention that these suggestions must be taken simply as a basis for discussion, D being free to modify his initial responses in whatever way he wishes. MACBETH belongs to the (up to now) restricted class of interactive approaches that follow a *learning paradigm* in decision aiding, allowing D to feel free in revising his judgments and, if he wants to go back anytime in his process of judging. If in the course of his cognitive process D decides to revise some of his initial judgments, making some modifications of category, the consistency of the new matrix of revised judgments will be tested again, thus re-starting the MACBETH interactive cycle of elaboration, modification and or validation of D absolute judgments of difference of attractiveness (see figure 4).

Let us add that, before applying program $Mc1$ to the judgments of D, it is very interesting, from a learning perspective, to start by testing the following condition (which follows immediately from (4)) called **semantic consistency condition**:

$$\forall a,b,c \in A \text{ with } aPb \text{ and } bPc \text{ and } \forall k,k' \in \{1,2,3,4,5,6\}, \text{ if } (a,b) \in C_k \text{ and } (b,c) \in C_{k'}, \text{ one must have } (a,c) \in C_{k''} \text{ with } k'' \geq \max\{k,k'\}.$$

This condition (which is implied by the cyclone condition) can easily be tested by verifying that:

In each row of D's matrix of judgments of difference of attractiveness the values $a_{i,j}$ do not decrease from left to right, and in each of its columns the values $a_{i,j}$ do not increase from top to bottom.

This is quite an easy practical test, which immediately shows up the judgments that can cause trouble when semantic consistency is not verified, thus creating learning conditions for launching an interactive discussion with D, where consistency can be reached by appropriately changing one or several of his initial judgments. This test is perfectly inserted in the constructive principles of our approach.

MACBETH's suggestion of a value scale: the program *Mc2*

As stated above, when c_{min} equals 0, one can be sure that there are real numbers $v(a)$ $(a \in A)$, s_1, s_2, s_3, s_4, s_5 and s_6 verifying **(3)** and **(4)**. The problem now is how to determine such numbers. Note that, when semantic consistency is verified, it can be interesting to give directly a suggestion of a numerical scale even when $c_{min} \neq 0$. Specially in multi-party or time-constrained decision processes, this can be a more pragmatic path for measurement than the more time-consuming revision of theoretically inconsistent judgments based upon the suggestions of (*Mc*3 and) *Mc*4. To give an answer to these two points, the MACBETH approach includes another linear program, called ***Mc*2**, the variables of which are $v(a)$ $(a \in A)$, s_1, s_2, s_3, s_4, s_5 and s_6: $\alpha(a,b)$ and $\delta(a,b)$ $[(a,b) \in C_6]$ and $\varepsilon(a,b)$ and $\eta(a,b)$ $[(a,b) \in C_1 \cup C_2 \cup \ldots \cup C_5]$

$$Min \left\{ \sum_{\substack{(a,b) \in C_k, \\ k \in \{1,2,3,4,5\}}} [\varepsilon(a,b) + \eta(a,b)] + \sum_{(a,b) \in C_6} \alpha(a,b) \right\}$$

s.t.

r0, *r1*, *r2*, *r3*, *r4* and *r5* (both with c replaced by c_{min})

r8) $\forall (a,b) \in C_6 : v(a) - v(b) = s_6 + 1 - \alpha(a,b) + \delta(a,b)$

r9) $\forall k \in \{1, 2, 3, 4, 5\}, \ \forall (a,b) \in C_k :$

$$v(a) - v(b) = (s_k + s_{k+1})/2 + \varepsilon(a,b) - \eta(a,b)$$

(see figure 3).

The formulation given to the objective-function of *Mc*2 results from our wish to arrive at numerical values for the difference of attractiveness of pairs assigned to the same category such that, if possible, these pairs are closer among themselves than pairs assigned to different categories. Note also that $\varepsilon(a,b).\eta(a,b) = 0$ in any basic solution of *Mc*2.

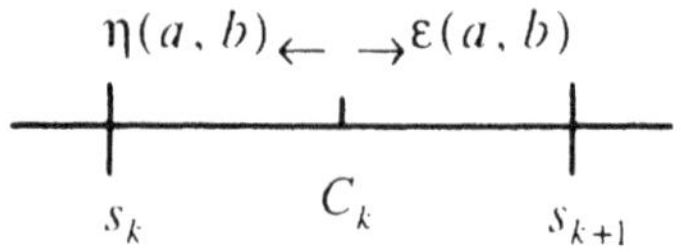

Figure 3 – Variables $\eta(a, b)$ and $\varepsilon(a, b)$

The scheme in figure 4 offers a graphic overview of the cyclic interactive process of the MACBETH approach. As figure 4 suggests, once a numerical scale is obtained (by the optimal solution of *Mc*2) the interaction with *D* can go on to the discussion of the cardinality of the scale, that is, to the validation of condition **(2)** (note that condition **(1)** is certainly satisfied). Probably, the most friendly way to ease this discussion will be to plot the numbers $v(a)$ $(a \in A)$ on an oriented axis, and than verify with *D* if the relative distances between points (actions) correspond to *D*'s relative differences of attractiveness. If so, the application $v: A \rightarrow \Re : a \rightarrow v(a)$ can be finally taken as an interval scale quantifying the attractiveness of the actions of *A* for *D*.

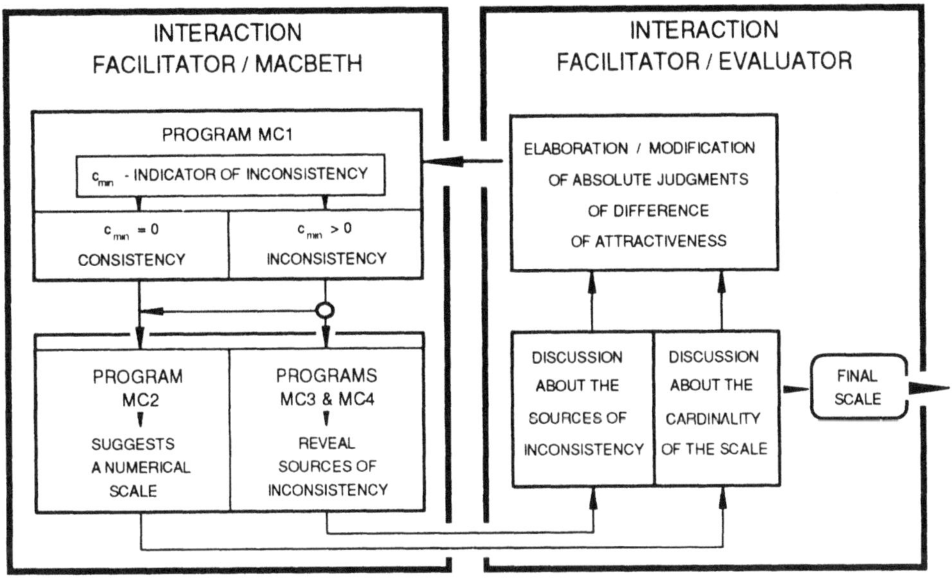

Figure 4 – Interactive scheme of MACBETH

Final comments

In Measurement Theory terminology, MACBETH is an interactive approach for mapping into a real scale various degrees of a property of the elements of a finite set *A*. The originality of MACBETH's questioning procedure is the possibility of establishing a constructive path towards cardinal measurement in both quantitative and substantive meaningful terms, avoiding the operational problems recognised as a weakness of other procedures. The use of the notion of semantic absolute judgments pays a key role here, and the simplicity, interactivity and constructiveness of our approach inserts it in the modern paradigms of decision aid.

In practice, the use of MACBETH strongly facilitates the interaction with the evaluators. As a matter of fact, in all the real-world decision aid cases where we have been applying our approach (see, for example, [Bana e Costa and Vansnick, 1995]) the evaluators not only felt comfortable in answering MACBETH's questions, emphasising the simplicity of the questioning procedure and the usefulness of MACBETH's suggestions to elaborate or revise judgments, but also they were very confident in the results of the interactive learning processes.

References

Bana e Costa, C.A., Vansnick, J.C. (1993), "Sur la quantification des jugements de valeur: L'approche MACBETH", *Cahier du LAMSADE*, 117, Université de Paris-Dauphine, Paris.

Bana e Costa, C.A., Vansnick, J.C. (1994a), "MACBETH – An interactive path towards the construction of cardinal value functions", *International Transactions in Operations Research*, 1, 4 (489–500).

Bana e Costa, C.A., Vansnick, J.C. (1994b), "A theoretical framework for Measuring Attractiveness by a Categorical Based Evaluation Technique (MACBETH)", J. Clímaco (ed.), *Proceedings of the XIth International Conference on MCDM*, Coimbra, Portugal, August 1994 (to appear).

Bana e Costa, C.A., Vansnick, J.C. (1995), "Applications of the MACBETH approach in the framework of an additive aggregation model", *Journal of Multi-Criteria Decision Analysis* (to appear).

Doignon, J.-P. (1987), "Threshold Representations of Multiple Semiorders", *SIAM Journal of Algebraic Discrete Methods*, 8, 1, 77–84.

Dyer, J.S., Sarin, R.K. (1979), "Measurable multiattribute value functions", *Operations Research*, 27, 4 (810–822).

Saaty, T.L. (1977), "A Scaling Method for Priorities in Hierarchical Structures", *Journal of Mathematical Psychology*, 15, 234–281.

Saaty, T.L. (1980), *The Analytic Hierarchy Process*, McGraw-Hill.

BUILDING ADDITIVE UTILITIES IN THE PRESENCE OF NON-MONOTONIC PREFERENCES

D.K. Despotis
University of Piraeus,
Dept. of Informatics, 80 Karaoli &
Dimitrioy St., 18534, Piraeus, Greece

C. Zopounidis
Technical University of Crete,
DSS Lab., University Campus,
73100, Chania, Greece

Abstract

Ordinal regression is used for modelling the problem of multiattribute choice in the presence of non-monotonic preferences. The main model presented in this paper estimates an additive utility function with non-monotonic partial utilities by analysing some ordinal inter-alternative information provided by the decision maker. Then the model is extended to fit in special conditions where intensity-of-preference information is available or concavity of the utility model is required.

Key words: Multicriteria Decision Making; Ordinal Regression

1. Introduction

The general problem of multiattribute choice deals with the exploration of a set of alternatives (courses of action, solutions, alternative decisions) $A=\{a_1, a_2, \ldots, a_m\}$, evaluated on a set of attributes $G=\{g_1, g_2, \ldots, g_n\}$, on the basis of some inter-alternative and/or inter-attribute preference information provided by the decision maker. Many normative decision methods have been proposed for modelling this problem

P.M. Pardalos et al. (eds.), Advances in Multicriteria Analysis, 101–114.

differing in various aspects, such asthe purpose they serve (choice, sorting or ranking of alternatives), the preference structures they assume, the information they require and the methodological background they use. The methods developed in the frame of multiattribute utility theory (MAUT) (commonly known as direct methods) and the regression-based methods (indirect methods) form two representative approaches to modelling the process of multiattribute choice. These approaches refer to a common axiomatic basis by assuming that the preference structure is a complete preorder structure thus represented by a utility function U. But, although they often share common utility-decomposition forms, they differ mainly in the methodology for assessing customised utility models from given analytical forms. In MAUT [cf. Keeney and Raiffa (1976)], the parameters of the utility model are estimated in a direct way on the basis of inter-attribute preference information obtained from the decision maker mainly in terms of judgements on attribute values. According to the regression-based approach [cf. Hammond et al. (1977), Jacquet-Lagreze and Siskos (1982), Oral and Kettani (1989)] the parameters are estimated by analyzing ordinal inter-alternative preference information provided by the decision maker.

A well known regression-based method is UTA (UTilite Additive) of Jacquet-Lagreze and Siskos (1982). The information that this method requires is a complete preorder of some reference alternatives ranked by the decision maker from the most to the least favoured with eventual ties according to his/her overall preferences. With this ordinal preference information as input, UTA fits ordinal utility functions of the additive form that best approximate the data by using a linear programming formulation and post-optimality analysis. The partial utilities are assumed non-linear, non-decreasing or non-increasing depending on the nature of the attributes, and are assessed in a piece wise linear form. A detailed review of the UTA method and its variations is given in Despotis et al. (1990). For a conceptually similar approach to the assessment of non-linear monotonic partial utilities the reader is referred to Oral and Kettani (1989).

An assumption commonly met in the literature is that of the monotonicity of preferences. It is usually assumed that the preference for an action evaluated on multiple attributes increases (or decreases) as the value of an attribute increases on the scale while the others remain constant. In the context of separable utility functions this means that the partial utilities are monotonic functions of the attributes. This assumption, although widely used, is sometimes not applicable to real world situations. For example, it can hardly be considered that one feels more and more comfortable as the temperature constantly increases or decreases in his sitting room. In a medical context, attributes such as level of sugar and cholesterol in blood are clear examples assuming non-monotonic preferences. In such type of attributes, preferences are monotonically increasing up to a predetermined desired level and monotonically decreasing for values exceeding that level.

One way to deal with attributes assuming non-monotonic preferences [cf. Keeney and Raiffa (1976)] is to divide the range of the attributes into intervals so that the preferences are monotonic in each interval and then treat each interval separately. Following this line of thinking and the conceptualization of non-linear partial utilities introduced in the UTA method, this paper presents an estimation model for the assessment of non-monotonic partial utilities. The formulation of the model treats both monotonic and non-monotonic partial utilities on a common basis and in that sense it can be viewed as a generalization of the UTA model. The model is developed in section 2 and illustrated with a numerical example in section 3. Finally, some possible extensions of the model are discussed in section 4.

2. Modelling non-monotonic preferences

As mentioned above, the problem of multiattribute choice involves a set $A=\{a_1, a_2, ..., a_m\}$ of alternatives and a set $G=\{g_1, g_2, ..., g_n\}$ of attributes which are used to evaluate the alternatives. Following the principles of multiattribute utility theory, it is assumed that the overall preferences on the alternatives and the preferences according to each attribute are complete preorder structures. This implies that there exists an unknown function U, the so called ordinal utility function, which implicitly forms the basis for the decision maker's choices. U is a composite real function

$$U = U(g_1, \ldots, g_n)$$

which aggregates all the attributes into a unidimensional index in a manner that if $(g_1(a), \ldots g_n(a))$ and $(g_1(b), \ldots, g_n(b))$ are the multiattribute evaluations of any two alternatives $a,b \in A$ and P and I denote respectively the strict preference and the indifference relations then

$$aPb \Leftrightarrow U(g_1(a), \ldots, g_n(a)) > U(g_1(b), \ldots, g_n(b))$$

$$aIb \Leftrightarrow U(g_1(a), \ldots, g_n(a)) = U(g_1(b), \ldots, g_n(b))$$

To assess overall utilities for the alternatives the utility model U is here assumed additively separable, thus having the form

$$U(a) = \sum_{i=1}^{n} U_i(g_i(a)) \tag{1}$$

where the U_i's are non-linear, non-monotonic partial utility functions. On the basis of the utility-decomposition model (1) the overall utility of each alternative $a_k \in A$ is approximated by

$$U'(a_k) = \sum_{i=1}^{n} U_i(g_i(a_k)) + \sigma^-(a_k) - \sigma^+(a_k)\ , \quad k=1, \ldots, m \tag{2}$$

where $\sigma^-(a_k)$ and $\sigma^+(a_k)$ are non-negative deviations representing respectively potential under achievement or over achievement with respect to the utility $U(a_k)$.

Let $G_i=[l_i, h_i]$ be the range of the attribute g_i, with l_i and h_i denoting respectively the lowest and the highest value of g_i, and d_i the most desirable value of g_i as determined by the decision maker. Since the partial utilities U_i are assumed non-monotonic, d_i might be at any point in the interval G_i and not necessarily at its extremities, that is $l_i \leq d_i \leq h_i$. Without loss of generality, it is assumed that each partial utility is non-decreasing from l_i to d_i and non-increasing from d_i to h_i, as shown in Figure 1.

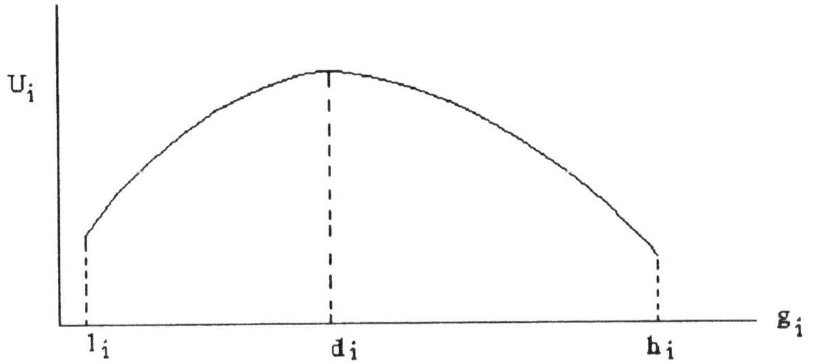

Figure 1: A non-monotonic partial utility function

To model non-linear, non-monotonic partial utilities, the range of each attribute is divided into two intervals $G_i^1=[l_i, d_i]$ and $G_i^2=[d_i, h_i]$. Then the former is discretized with p_i equally spaced breakpoints g_i^j, $j=1,...,p_i$ with $g_i^1=l_i$ and $g_i^{p_i}=d_i$ and the latter is similarly discretized with q_i+1 breakpoints labelled as $g_i^{p_i}$, $g_i^{p_i+1}$, ..., $g_i^{p_i+q_i}$ with $g_i^{p_i+q_i}=h_i$. The parameters p_i and q_i are estimated according to the dispersion of the attribute scores in G_i, in a manner that each sub-interval $[g_i^j, g_i^{j+1}]$ contains at least one value $g_i(a_k)$. Then for every value $g_i(a_k)$ there is exactly one $j \in \{1,..., p_i+q_i\}$, depending on i and k (j(i,k) say) such that $g_i(a_k) \in [g_i^{j(i,k)}, g_i^{j(i,k)+1}]$. Hence the partial utility $U_i(g_i(a_k))$ is approximated by linear interpolation

$$U_i(g_i(a_k)) = t_i^k U_i(g_i^{j(i,k)}) + (1 - t_i^k) U_i(g_i^{j(i,k)+1}) \tag{3}$$

with

$$t_i^k = \frac{g_i^{j(i,k)+1} - g_i(a_k)}{g_i^{j(i,k)+1} - g_i^{j(i,k)}}$$

In that manner the partial utility functions U_i are assessed in a piece-wise linear fashion and the overall utility of every alternative a_k is then given in terms of the partial utilities of the breakpoints as follows:

$$U'(a_k) = \sum_{i=1}^{n} \{t_i^k U_i(g_i^{j(i,k)}) + (1 - t_i^k) U_i(g_i^{j(i,k)+1})\} + \sigma^-(a_k) - \sigma^+(a_k) \quad (4)$$

Setting

$$U_i(g_i^{j+1}) - U_i(g_i^j) = \begin{cases} w_{ij} & if\ 1 \le j \le p_i - 1 \\ -w_{ij} & if\ p_i \le j \le p_i + q_i - 1 \end{cases}$$

for every pair of successive breakpoints g_i^j, g_i^{j+1}, the partial utilities of the breakpoints are expressed in terms of the variables w_{ij} through the relations

$$U_i(g_i^j) = \begin{cases} \sum_{d=0}^{j-1} w_{id} & if\ 1 \le j \le p_i \\ \sum_{d=0}^{p_i-1} w_{id} - \sum_{d=p_i}^{j-1} w_{id} & if\ p_i + 1 \le j \le p_i + q_i \end{cases} \quad (5)$$

and the partial utilities of the alternatives are written

$$U_i(g_i(a_k)) = \begin{cases} (\sum_{d=0}^{j(i,k)-1} w_{id}) + (1 - t_i^k) w_{i,j(i,k)} & if\ 1 \le j(i,k) \le p_i - 1 \\ \sum_{d=0}^{p_i-1} w_{id} - (\sum_{d=p_i}^{j(i,k)-1} w_{id}) - (1 - t_i^k) w_{i,j(i,k)} & if\ p_i \le j(i,k) \le p_i + q_i - 1 \end{cases} \quad (6)$$

From (6) it is derived that the partial utilities at the extremities l_i and h_i of each interval G_i are

$$U_i(l_i) = w_{i0}, \quad U_i(h_i) = \sum_{d=0}^{p_i-1} w_{id} - \sum_{d=p_i}^{p_i+q_i-1} w_{id}$$

The preference information required from the decision maker to assess the utility model is an ordinal ranking of the alternatives in A. To simplify the notation, A is assumed to be ordered by S=P∪I, that is sequenced according to the decision maker's subjective preferences, a_1 and a_m being respectively the most and the least favoured alternatives. Thus for every pair of successive alternatives a_k, $a_{k+1} \in A$ one of the following holds:

$$\begin{aligned} a_k P a_{k+1} &\Leftrightarrow U'(a_k) - U'(a_{k+1}) \geq \delta \\ a_k I a_{k+1} &\Leftrightarrow U'(a_k) - U'(a_{k+1}) = 0 \end{aligned} \qquad (7)$$

δ being a small positive number introduced to discriminate significantly two successive equivalence classes of S.

From the expressions (4)-(6) it follows that for the estimation of the overall utilities of the alternatives a_k, k=1,..., m it is sufficient to estimate the variables w_{ij}, i=1,..., n, j=0,..., p_i+q_i-1, taking into account the conditions (7) which reflect the subjective preferences of the decision maker with respect to all $a_k \in A$. The estimation of w_{ij}, and thus of a piece-wise linear utility function, is made, at a minimum mean deviation from the ordinal ranking provided by the decision maker, by means of the following linear programming model:

$$\min \left\{ \sum_{k=1}^{m} \sigma^{+}(a_k) + \sigma^{-}(a_k) \right\}$$

subject to

$$\begin{aligned} & U'(a_k) - U'(a_{k+1}) \geq \delta \quad if \quad a_k P a_{k+1} \\ & U'(a_k) - U'(a_{k+1}) = 0 \quad if \quad a_k I a_{k+1}, \quad k = 1, \ldots, m-1 \end{aligned} \qquad \text{(P1)}$$

$$\sum_{d=0}^{p_i-1} w_{id} - \sum_{d=p_i}^{p_i+q_i-1} w_{id} \geq 0, \quad i=1,\ldots, n$$

$$w_{ij} \geq 0, \sigma^{+}(a_k) \geq 0, \sigma^{-}(a_k) \geq 0, i=1,\ldots, n, j=0,\ldots, p_i+q_i-1, k=1,\ldots, m$$

In the model P1, the most preferred values d_i of one or more attributes can be replaced by intervals in which the partial utility functions attain their highest values and are constant. This is the case illustrated with a numerical example in the next section.

3. Numerical example

Consider the case of fifteen firms evaluated on the following attributes (financial ratios) giving a picture of their financial situation and performance (cf. Zopounidis, 1990): (1) *cash to total assets*; (2) *long term debt and stockholder's equity to fixed assets*; (3) *total liabilities to total assets*. The firms have the attribute values as given in Table 1.

Table 1: Multiattribute evaluation of the firms

Firms	Attributes		
	Cash/Total assets (g_1)	Long term Debt +Stockholder's equity/Fixed assets (g_2)	Total liabilities/ Total assets (g_3)
F1	3.80	2.40	60.70
F2	5.84	1.96	63.70
F3	0.04	1.14	64.26
F4	4.89	2.92	55.04
F5	0.57	1.72	64.70
F6	16.70	2.32	53.29
F7	3.16	4.10	23.90
F8	25.42	3.35	59.03
F9	17.99	1.34	73.84
F10	3.98	3.26	84.95
F11	0.76	2.74	84.44
F12	24.16	2.83	70.51
F13	2.53	2.54	81.05
F14	35.06	9.56	61.08
F15	0.72	0.97	99.67

The first ratio indicates the cash position of the firm, that is it's capacity to cope with the operating expenditures and other cash obligations when and as they fall due. From a theoretical point of view, the higher the ratio the higher the cash resources available to the firm. However in practice, if a firm possesses sufficient resources and can not manage them properly, it loses investment opportunities liable to bring in increased profits; it mismanages finances keeping money reserves that remain inert. To model this situation the attribute values ranging from 3% to 5% are taken to be the most satisfactory. Values bellow 3% and above 5% are considered less satisfactory.

The second ratio indicates the financing of fixed assets by stockholder's equity or long term debt. This is a working capital ratio. A value less than 1 means that the firm does not have long term resources to finance long term investment needs. However, a high-valued ratio could mean that the firm is compromising fixed assets (disinvestment) because of old age or old technology, and does not invest any more. Modelling this situation, values ranging from 1 to 2.5 are taken to be satisfactory. Scores bellow 1 and above 2.5 are considered less satisfactory.

The third ratio indicates the debt capacity of the firm. The higher this ratio the higher the proportion of assets financed by non-shareholder parties. In practice however, as the financial analysts believe, once the ratio exceeds 67%, the firm is regarded as highly indebted and in increased financial risk. On the other hand, a low value means that the firm is not taking enough advantage of the financial leverage (financing by loan) to improve its return on equity. Thus it is assumed here that the most satisfactory values for the third attribute range between 50% and 65%.

The decision maker evaluates the fifteen firms by taking into account their attribute values and ranks the firms in the following order of preference:

1: {F1, F4}; 2: {F2}; 3: {F5}; 4: {F6}; 5: {F13}; 6: {F3}; 7: {F8};
8: {F10}; 9: {F12, F7}; 10: {F9}; 11: {F11}; 12: {F14}; 13: {F15}

Applying model P1 to the data of Table 1, with $p_1=3$, $q_1=5$, $p_2=2$, $q_2=3$, $p_3=2$, $q_3=3$, $\delta=0.005$, and then solving the resulting optimization problem the following solution is obtained: $w_{10}=0.065$, $w_{11}=0.097$, $w_{12}=0$, $w_{13}=0.087$, $w_{14}=0$, $w_{15}=0.021$, $w_{16}=0.054$, $w_{20}=0$, $w_{21}=0.007$, $w_{22}=0$, $w_{23}=0$, $w_{30}=0$, $w_{31}=0.121$, $w_{32}=0.107$, $w_{33}=0$, with $F^*=0$. The optimal additive utility model is given graphically in Figure 2 in terms of the partial utility curves of the attributes. The overall utilities of the alternatives are: $U_1=0.290$, $U_2=0.280$, $U_3=0.193$, $U_4=0.290$, $U_5=0.228$, $U_6=0.203$, $U_7=0.168$, $U_8=0.188$, $U_9=0.158$, $U_{10}=0.183$, $U_{11}=0.134$, $U_{12}=0.168$, $U_{13}=0.198$, $U_{14}=0.128$, $U_{15}=0.124$.

The utility model is estimated at a null deviation from the decision maker's subjective preference order, that is a complete agreement between the input ranking and that derived by the model is achieved.

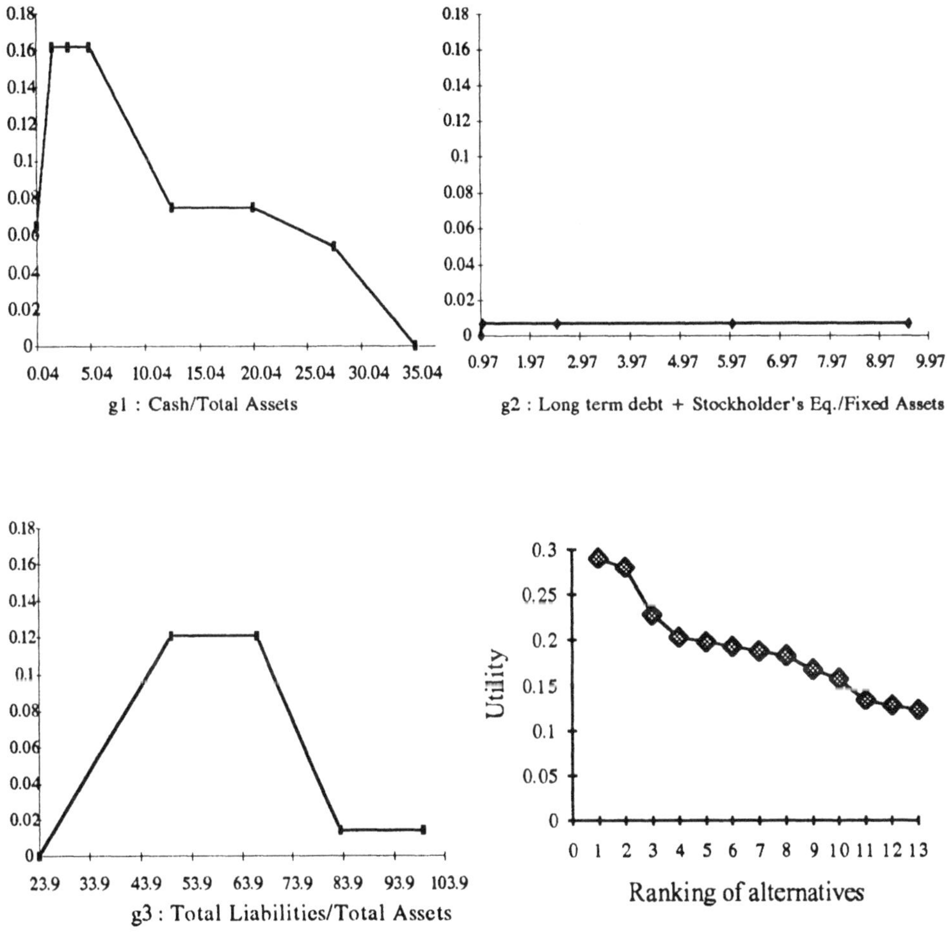

Figure 2: Partial and overall utility curves

4. Extensions and remarks

To implement the model P1 to a multiattribute decision situation the decision maker uses an ordinal scale and on the basis of his/her implicit preference model ranks order the alternatives from the most to the least favoured, without expressing intensity of preferences. Moreover, the decision maker defines the most favoured values of the attributes, either as specific values or as intervals, since, in the presence of non-monotonic preferences, these values do no more constitute an objective reality deriving from the data. However, the model treats both monotonic and non-monotonic preferences on a common basis. In fact, if $d_i=h_i$ or $d_i=l_i$ for some attributes, then for these attributes the most favoured values are self-defined and the corresponding partial utilities are estimated as monotonic functions, non-decreasing or non-increasing respectively.

Under different circumstances, if the decision maker ranks the alternatives by using an explicit overall index I (for example ranks alternative products according to their market shares) or uses a ratio scale to express intensity of preferences [cf. Oral and Kettani (1989)] , the utility estimation model should be written as follows:

$$\min \ \{\sum_{k=1}^{m} \sigma^{+}(a_k) + \sigma^{-}(a_k)\}$$

subject to

$$U'(a_k) - U'(a_{k+1}) = I_k - I_{k+1} \ , \ k=1,\dots,m-1 \qquad \text{(P2)}$$

$$\sum_{d=0}^{p_i-1} w_{id} - \sum_{d=p_i}^{p_i+q_i-1} w_{id} \geq 0, \ i=1,\dots,n$$

$$w_{ij} \geq 0, \sigma^{+}(a_k) \geq 0, \sigma^{-}(a_k) \geq 0, i=1,\dots,n, \ j=0,\dots,p_i+q_i-1, \ k=1,\dots,m$$

The first set of constraints in P2 states not only the succession of the alternatives in the preference ranking (as in P1) but also that the difference between the overall utilities of any two successive alternatives in the ranking should be consistent with the difference of their evaluations on the ratio scale.

Under special conditions, it is necessary the estimated utility function to possess some additional properties, such as concavity. This is the case in the context of multiobjective mathematical programming, where the utility function must be maximized to estimate a solution of global maximum. To assess a concave utility function in the presence of non-monotonic preferences the model P1 could be adjusted as follows:

$$\min \{\sum_{k=1}^{m} \sigma^{+}(a_k) + \sigma^{-}(a_k)\}$$

subject to

$$\begin{aligned} U'(a_k) - U'(a_{k+1}) &\geq \delta \quad if \quad a_k P a_{k+1} \\ U'(a_k) - U'(a_{k+1}) &= 0 \quad if \quad a_k I a_{k+1}, \quad k = 1, \ldots, m-1 \end{aligned}$$

$$\sum_{d=0}^{p_i - 1} w_{id} - \sum_{d=p_i}^{p_i + q_i - 1} w_{id} \geq 0, \; i = 1, \ldots, n \qquad \text{(P3)}$$

$$\begin{aligned} w_{i,j-1} - w_{ij} &\geq 0, \quad i = 1, \ldots, n, \; j = 2, \ldots, p_i - 1 \\ w_{i,j-1} - w_{ij} &\leq 0, \quad i = 1, \ldots, n, \; j = p_i + 1, \ldots, p_i + q_i - 1 \end{aligned}$$

$$w_{ij} \geq 0, \sigma^{+}(a_k) \geq 0, \sigma^{-}(a_k) \geq 0, i = 1, \ldots, n, \; j = 0, \ldots, p_i + q_i - 1, \; k = 1, \ldots, m$$

The model P3 is derived from P1 by appending a new set of constraints assuring that the estimated partial utility functions are concave.

References

Despotis, D.K., D. Yannacopoulos, C. Zopounidis (1990) A review of the UTA multicriteria method and some improvements, **Foundations of Computing and Decision Sciences**, 15, 2, 63-76.

Hammond, K.R., R.L. Cook, L. Adelman (1977) POLICY: An aid for decision making and international communication, **Columbia Journal of World Business**, 12, 79-93.

Jacquet-Lagreze, E., J. Siskos (1982) Assessing a set of additive utility functions for multicriteria decision making. The UTA method, **European Journal of Operational Research**, 10, 151-164.

Keeney, R. L., H. Raiffa (1976) Decisions with multiple objectives: Preferences and value trade-offs (**John Wiley and Sons, New York**).

Oral, M., O. Kettani (1989) Modelling the process of multiattribute choice, **Journal of Operational Research Society**, 40, 281-291.

Siskos, J., D. Yannacopoulos (1985) UTASTAR - An ordinal regression method for building additive value functions, **Investigacao Operational**, 5, 39-53.

Zopounidis, C. (1990) La gestion du capital-risque (**Editions Economica, Paris**).

IV. MULTI-OBJECTIVE OPTIMIZATION

THE IDEAL-DEGRADATION PROCEDURE:

Searching for Vector Equilibria

Milan Zeleny

Graduate School of Business Administration, Fordham University at Lincoln Center, New York, New York 10023, U.S.A.

The Ideal Degradation Procedure (IDP) is designed to solve multiple criteria problems without scalarization, i. e., in the vector-maximum sense. Scalarization reduces multiple criteria into a single aggregate superfunction, thus defeating the very purpose and all advantages of explicitly considering multidimensionality. Very few OR/MS analysts would propose "collapsing" problem constraints into a single "superconstraint". Yet, multiple criteria are being freely scalarized even when they express the same economic variables as constraints.

Traditional *ideal solution* of compromise programming represents the most preferred solution under all and any well behaved scalar (utility) functions. It is therefore, like the set of nondominated solutions, a vector-maximum concept. Because ideal solutions (and their superiors) are typically infeasible, short of making them feasible through design, a compromise must be reached: the ideal has to be degraded (downgraded, reduced). This ideal degradation towards compromise must be pursued without scalarization (e. g., distance function), but in a vectorwise, dimensionality preserving mode. The proposed IDP is designed to reformulate the ideal towards acceptable compromise according to differential weights of criteria importance.

Introduction

John von Neumann and Oskar Morgenstern had the following to say about the *vector optimization* problem [2]:

> This is certainly no maximum problem, but a peculiar and disconcerting mixture of several conflicting maximum problems ... This kind of problem is nowhere dealt with in classical mathematics. We emphasize at the risk

P.M. Pardalos et al. (eds.), Advances in Multicriteria Analysis, 117–127.

> of being pedantic that this is no conditional maximum problem, no problem of the calculus of variation, of functional analysis, etc. It arises in full clarity, even in the most "elementary" situations, e. g., when all variables can assume only a finite number of values.

A vector optimization problem [3] of Kuhn and Tucker, 1951, arises when two or more (K) scalar-valued objective functions (or criteria) are to be maximized over a set of feasible solutions:

$$opt\ \{f(x) \mid x \in X\}, \tag{1}$$

where $X = \Re^N$ and $f(x) = (f_1(x), \ldots, f_K(x))$. A feasible solution x" is nondominated with respect to X and $f(x)$ if and only if there exists no $x' \in X$ such that $f(x') \geq f(x'')$ and $f(x') \neq f(x'')$. Solving a vector optimum problem means identifying all x", i. e. the set N(X), which are nondominated with respect to X and $f(x)$.

Each kth component scalar-optimum problem,

$$opt^+\{f_k(x) \mid x \in X\},\ k = 1, \ldots, K \tag{2}$$

has an optimal solution x^{k+} with $f^{k+} = f_k(x^{k+})$. The vector $f^+ = (f^{1+}, \ldots, f^{K+})$ represents the ideal point of $f(x)$ with respect to X. If $x^{k+} = x^+$, k = 1. ..., K, the vector optimum problem has a *perfect solution* $f^+ = f(x^+)$. This case excepted, typically no feasible x yields vector f^+. It is of course possible to redefine (or "reshape") X so that perfect solution is enclosed and the vector optimization problem effectively resolved [4]. We shall consider both cases: i. e., X fixed and flexible, fully or partially.

The perfect solution f^+ is mirrored by the lowest, separately acceptable values achieved by anti-optimizing over the same nondominated set N(X) of X. Each component scalar anti-optimum problem,

$$opt^-\{f_k(x) \mid x \in N(X)\},\ k = 1, \ldots, K \tag{3}$$

has an anti-optimal solution x^{k-} with $f^{k-} = f_k(x^{k-})$. The vector $f^- = (f^{1-}, \ldots, f^{K-})$ represents the anti-ideal point of $f(x)$ with respect to N(X). If $x^{k-} = x^-$, k = 1. ..., K, the vector optimum problem has an *imperfect solution* $f^{-} = f(x^{-})$.

The imperfect solution can not acceptable as a whole (it is necessarily dominated) although its individual component scores are acceptable separately (as they could become part of N(X)).

Performance scores of perfect and imperfect solutions identify the ranges of usable values $[f^{k-}, f^{k+}]$, k = 1. ..., K, defined over N(X). These ranges provide realizable potentials for change or tradeoffs for each criterion, as in the earlier technique IDEA [7].

Barring several algorithmic results in identifying and interactively reducing (or filtering) the set of nondominated solutions, the above summarizes most of the advances made by MCDM in the area of vector optimization [6].

Even though MCDM formally commenced from vector *optimization*, it has not passed beyond the above initiatory definitions and calculations. The vector optimization problem has been directly and without sufficient justification transformed into an aggregate *scalar* optimization problem and most research concentrated on determining the multiplier "weights of importance" for component criteria functions comprising the aggregate scalar function.

Vector optimization problem (1) has been conveniently replaced by a scalar maximization problem:

$$max \{U(f(x)) \mid x \in X\}, \tag{4}$$

or by

$$max \{U(w, f(x)) = w_1 f_1(x) + ... + w_K f_K(x) \mid x \in X\}, \tag{5}$$

or by some other aggregate utility-function variation. Multiple criteria problem has mutated into a single criterion problem. Yet the name, Multiple Criteria Decision Making (MCDM), persists from the first MCDM conference in South Carolina in 1973.

Weights of Importance

The U(w, f(x)) is a single-criterion aggregate structure that overrules individual criteria and renders their weights of importance, w_k, k = 1, ..., K, either meaningless or simple normalizing multipliers. The larger the weight w_k, the larger the contribution of criterion score to the overall aggregate of U(w, f(x)).

Why should the concept of criterion importance be related to or derived from the overall performance of aggregate superfunction U(w, f(x)) is not trivial to answer.

Maximization of the aggregate is a separate criterion which has to be justified. It can enhance or overrule any *a priori* expressed estimates of other criteria importance. Whether or not to add an aggregate (or even replace other criteria by it) is a value judgment stemming from the notion that collective performance is more important than individual performances. This is not a generally acceptable concept: It is not at all clear that if I want to maximize criteria of mileage *and* quality, I should settle for maximizing some aggregate (mileage + quality) instead, without regard to their individual achievements.

Consider a simple example [5]. Let w_k = 0.2, k=1, ..., 5 be equal weights of importance of five different criteria. In selecting among automobiles we may, for example, consider comfort, quality, price, reliability and mileage as equally important criteria. Comparing two available automobiles, A and B, assume that the following criteria scores were recorded:

A: (17; 1; 1; 1; 1) U(A) = 4.2

B: (4; 4; 4; 4; 4) U(B) = 4.0

The weighted aggregates U(w, f(x)) achieve values 4.2 and 4.0 for A and B respectively. Any aggregate maximizer would choose automobile A with U = 4.2, *ceteris paribus*. He would end up with an extravagantly comfortable car (score 17) of inferior quality, reliability and mileage at very high price (scores 1). The whole point of specifying *a priori* equal importance for all five criteria has been missed. The actual outcome makes one criterion (comfort) overwhelmingly important and all other criteria disagreeably unimportant. There is no sense of equal-weight balance, harmony or equilibrium: maximization of U overrides decision maker's preferences.

Clearly, free decision maker does not typically seek to maximize any aggregate, superfunction or collective performance. Free agents search for a specific (e. g., equally weighted) mix of real criteria performance scores, ideal if possible, closest to the ideal if necessary. Searching for close-to-ideal equilibrium is his primary concern.

The *ex post* fact that some U can be formally maximized at decision maker's equilibrium choice is secondary, a result of conventional tautology: That which maximizes U is preferred; that which is preferred maximizes U.

There are two basic decision-making assumptions:

1. Single criterion. Humans maximize superfunction U, including the case of w_k = 1 for one $f_k(x)$. The weights of importance are normalization multipliers or discount factors which bring the achieved criteria scores differentially into the weighted total. Individual performances are less important than the overall performance, collective stands above the individual.
2. Multiple criteria. Humans do not maximize collective superfunction but search for balanced or equilibrium performance, close to perfect solution in accordance to the expressed weights of individual criteria importance. Individual performances are more important than the overall performance, collective is secondary and individual is primary.

Each decision situation is characterized by a specific context of feasible alternatives. This context specifies currently available maximal and minimal performances (perfect and imperfect solutions) attainable with respect to all criteria. Humans expect performance scores of important criteria to be close to such maxima (or far from minima), while allowing less important criteria to score further away from maxima (or closer to minima). If all criteria are equally important, then their individual performance scores are expected to be proportionately equidistant from the corresponding maxima.

If there is only one single criterion, say Max U(w, $f(x)$), then there is no problem: its constrained maximum (and ideal) value or score is attainable by definition.

Degradation Procedure

Perfect solution f^+ offers well balanced, equilibrium [1, 8] "profile" (vector) of scores to be accepted by all rational decision makers. If there is no feasible x for which $f(x) = f^+$, the perfect solution f^+ has to be degraded closer to X while maintaining its vector contour. Final optimal solution, although degraded, resembles (or approximates) the perfect solution *as closely as possible.*

It is now appropriate to draw distinction between resemblance (or approximation) and *distance*. Our notion of resemblance refers to vector *profile matching*, measured by weights-derived rates of change. Distance is typically measured by algebraic *aggregate formulas*, ranging from assorted Lp-norms to entropy measures.

In order to normalize all criteria scores within their own context of measurement, we employ the following unidimensional distance measure:

$$d_k(x) = \frac{f_{k+} - f_k(x)}{f_{k+} - f_{k-}} \qquad \text{for all } k = 1, \ldots, K \tag{6}$$

where $0 \leq d_k(x) \leq 1$ for all $x \in N(X)$, i. e., $d_k(x^{k+}) = 0$ and $d_k(x^{k-}) = 1$ for k = 1, ..., K.

The degradation process then proceeds from the perfect vector, with distance $d^+ = (0, 0, \ldots, 0)$, towards the imperfect vector, with distance $d^- = (1, 1, \ldots, 1)$, according to the vector of relative weights of importance $w = (w_1, w_2, \ldots, w_K)$:

$$d(x) = d^+ + w \cdot s,$$

where $s = (s_1, s_2, \ldots, s_K)$ is the degradation step size. Initially, $s_1 = s_2 = \ldots = s_K$.

Example. Let perfect vector $f^+ = (300; 20; 1100; 5; 60)$ and assume that imperfect vector $f^- = (200; 18; 600; 1; 30)$. The ranges of the five criteria are (100; 2; 500; 4; 30), indicating also the maximum (100%) step sizes. Let us assume that all the step sizes are equally important according to the weights of criteria importance w = (.2, .2, .2, .2, .2). So, if we choose the step size to be 50%, it shall remain 50% even after the weighted normalization. The 50%-degraded, perfect matching vector is (50; 1; 250; 2; 15).

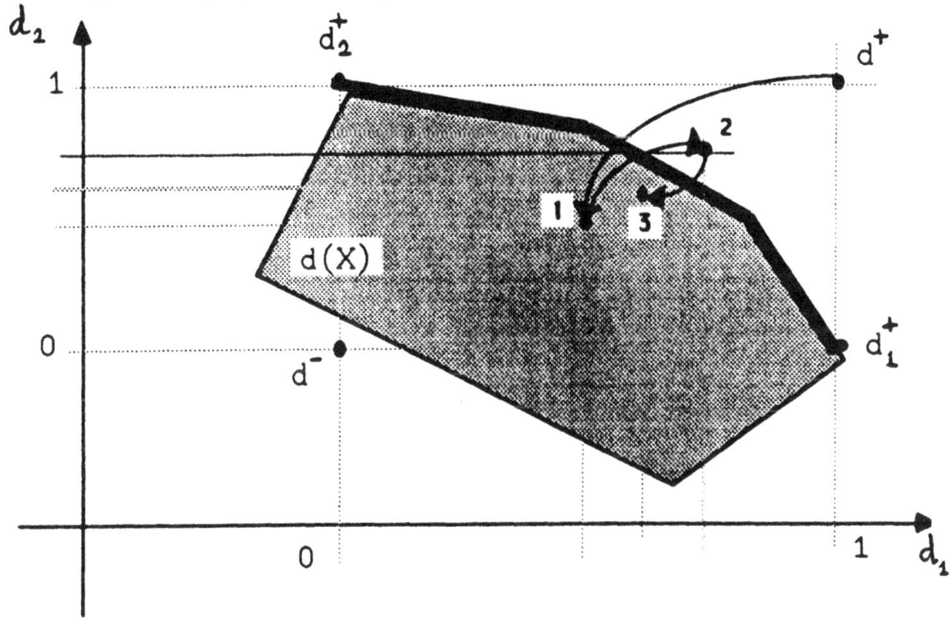

Fig. 1. Example of the Degradation Procedure

The degradation procedure ends as soon as the first $x \in N(X)$ is identified. In Fig. 1., the process is outlined for two equally important criteria in their normalized space. This means that unequal-size criteria ranges [0, 1] are being halved in making the successive degradation moves. Iteration points start at d^+ and are successively labeled as 1, 2, 3 and so on.

Perfect solution d^+ is well balanced and the purpose of its successive degradation is to maintain its perfect balance with respect to relative weights of criteria importance. Nondominated solution cannot always be reached by maintaining perfect balance: sometimes either tradeoffs or changes in the feasible set will be necessary.

Linear-Programming Formulations

Multiobjective linear-programming formulations are straightforward:

$$\max Cx \quad \text{s.t.} \quad Ax - b \leq 0, \quad pb \leq B, \quad x \geq 0, \tag{8}$$

where $Z = Cx = \sum_{kj} c_{kj} x_{kj}$, $k = 1, \ldots, K$, are K objective functions z_k to be maximized simultaneously. Clearly, $pAx \leq B$ follows from (8), and thus defining vector of unit cost $V = [V_1, \ldots, V_n] = pA$, we can rewrite (8) as:

$$\max Z = Cx \quad \text{s.t.} \quad Vx \leq B, \quad x \geq 0. \tag{9}$$

Using the methodology of *De novo programming* [4], we can solve problem (9), for x and b, with respect to each objective function z_k separately. Let $z^{k^*} = \max z_k$, $k = 1, \ldots, K$, subject to the constraints (9). Let vector $Z^* = (z^{1^*}, \ldots, z^{K^*})$ denote the multicriteria performance of the ideal design relative to given B. Obviously, Z^* must be unattainable for a given budget level B.

Observe that to each of the K values of Z^*, there is a corresponding z^{k^*}-optimal resource portfolio b^*. Vector Z^* represents the *metaoptimal performance*: we can find its corresponding pattern x^* and b^* by solving the following problem:

$$\min B = Vx \quad \text{s.t.} \quad Cx = Z^*, \quad x \geq 0 \tag{10}$$

Solving problem (10) identifies the minimum budget B^* at which the metaoptimum performance Z^* can be realized through x^* and b^*. Solution to (10), B^* and x^*, is designated as *metaoptimum solution.*

Observe that $B^* \geq B$, i. e., metaoptimum budget B^* must exceed any given budget B we have started from. If we wish to enforce any particular budget level B, we use the *optimum-path ratio* ***r***:

$$r = B/B^* \tag{11}$$

and establish the final solution as $\mathbf{x} = \boldsymbol{r}x^*$, $\mathbf{b} = \boldsymbol{r}b^*$ and $\mathbf{Z} = \boldsymbol{r}Z^*$.

The optimum-path ratio ***r*** provides an efficient tool for virtually instantaneous optimal redesign of even large-scale *linear* systems.

The following numerical example is adapted from Zeleny [4]:

$$\begin{aligned}
\text{Maximize} \quad & z_1 = 50x_1 + 100x_2 + 17.5x_3 \\
& z_2 = 92x_1 + 75x_2 + 50x_3 \\
& z_3 = 25x_1 + 100x_2 + 75x_3 \\
\text{subject to} \quad & \\
& 12x_1 + 17x_2 \leq b_1 \\
& 3x_1 + 9x_2 + 8x_3 \leq b_2 \\
& 10x_1 + 13x_2 + 15x_3 \leq b_3 \\
& 6x_1 + 16x_3 \leq b_4 \\
& x12x_2 + 7x_3 \leq b_5 \\
& 9.5x_1 + 9.5x_2 + 4x_3 \leq b_6
\end{aligned} \tag{12}$$

We assume that the weights of importance of objective functions z1, z2, z3 are identical, w1 = w2 = w3, for the sake of simplicity. Problem (12) calls for identifying optimal resource levels of b1 through b6 when resource unit prices are given as p1 = 0.75, p2 = 0.60, p3 = 0.35, p4 = 0.50, p5 = 1.15 and p6 = 0.65 respectively. Let B = 4658.75.

We calculate Z^* = (10916.813, 18257.933, 12174.433) with respect to given B (= 4658.75). The feasibility of Z^* can be only assured by metaoptimum solution x^* = (131.341, 29.683, 78.976) at the cost of B^* = 6616.5631.
Because the optimal-path ratio ***r*** = 4658.75/6616.5631 = 70.41%, the resulting **x**

= (92.48, 20.90, 55.61) and **Z** = (7686.87, 12855.89, 8572.40)

It follows that optimal portfolio **b**, with respect to B = 4658.75, can be calculated by substituting **x** in the constraints of (12). We obtain:

$$\begin{aligned} b_1 &= 1465.06 \\ b_2 &= 910.42 \\ b_3 &= 2030.65 \\ b_4 &= 1444.64 \\ b_5 &= 640.07 \\ b_6 &= 1299.55 \end{aligned} \tag{13}$$

IIf we spend precisely B = 4658.8825 (approx. 4658.75) then the optimum portfolio of resources to be acquired at current market prices is (13), allowing us to produce **x** and realize **Z** in criteria performance.

References

1. Galperin, E. A., "Nonscalarized Multiobjective Global Optimization," *Journal of Optimization Theory and Applications*, 75(1992)1, pp. 69 - 86.

2. Von Neumann, J. and O. Morgenstern, *Theory of Games and Economic Behavior*, Third edition, Princeton University Press, Princeton, N. J., 1953, pp. 10 - 11.

3. Kuhn, H. W. and A. W. Tucker, "Nonlinear Programming," in: J. Neyman (ed.), *Proceedings of the Second Berkeley Symposium on Mathematical Statistics and Probability*, University of California Press, Berkeley, CA, 1951, p. 488.

4. Zeleny, M., "Optimizing Given Systems vs. Designing Optimal Systems: The *De Novo* Programming Approach," *General Systems*, 17(1990)4, pp. 295 - 307.

5. Zeleny, M., "Measuring Criteria: Weights of Importance," *Human Systems Management*, 10(1991)4, pp. 237 - 238.

6. Zeleny, M.,"An Essay Into a Philosophy of MCDM: A Way of Thinking or Another Algorithm?," Invited Essay, *Computers and Operations Research*, 19(1992)7, pp. 563 - 566.

7. Zeleny, M., "Interactive Decision evolution Aid (IDEA)," in: *Multiple Criteria Decision Making*, McGraw-Hill, New York, 1982, pp. 369 - 373.

8. Zeleny, M., "MCDM: Return to Vector Optimization," *Proceedings of the 10th International Conference on MCDM*, Taipei, Taiwan, July 19 - 24, 1992, pp. 7 - 14.

THE LIGHT BEAM SEARCH - OUTRANKING BASED INTERACTIVE PROCEDURE FOR MULTIPLE-OBJECTIVE MATHEMATICAL PROGRAMMING

Andrzej JASZKIEWICZ Roman SŁOWIŃSKI

Institute of Computing Science
Poznań University of Technology
60-965 Poznań, Poland

Abstract. An interactive procedure for multiple-objective analysis of linear and non-linear programs is presented. At the decision phase of the procedure, a sample of points, composed of the current point and a number of alternative proposals, is presented to the decision maker (DM). The sample is constructed to ensure a relatively easy evaluation of the sample by the DM. To this end we use an outranking relation as a local preference model in a neighbourhood of the current point. The outranking relation is used to define a sub-region of the non-dominated set the sample presented to the DM comes from. The DM has two possibilities, or degrees of freedom, to move from one sub-region to another which better fits his/her preferences. The first possibility consists in specifying a new reference point which is then projected onto the non-dominated set in order to find a better non-dominated point. The second possibility consists in shifting the current point to a selected point from the sub-region. In both cases, a new sub-region is defined around the updated current point. This technique can be compared to projecting a focused beam of light from a spotlight in the reference point onto the non-dominated set; the highlighted sub-region changes when either the reference point or the point of interest in the non-dominated set are changed.

1. Introduction

In the general case of multiple-objective linear and non-linear mathematical programming, the decision problem consists in selecting the best compromise solution from an infinite multi-dimensional set of non-dominated alternatives. It is commonly acknowledged that interactive procedures are very effective in searching over the non-dominated set for the best compromise. Procedures of this type are characterized by phases of decision alternating with phases of computation. At each computation phase, a solution, or a subset of solutions, is generated for examination in the decision phase. As a

P.M. Pardalos et al. (eds.), Advances in Multicriteria Analysis, 129–146.

result of the examination, the DM inputs some preferential information which intends to improve the proposal(s) generated in the next computation phase.

A number of interactive procedures that present to the DM one point only at each iteration, has been proposed. This class of methods includes such well-known representatives like: STEM (Benayoun et al., 1971), interactive goal programming (see e.g. Lee and Shim, 1986), the reference point method (Wierzbicki, 1980) and Pareto Race (Korhonen and Wallenius, 1988). We claim, however, that presentation of one solution at each iteration does not give the DM a possibility to learn much about the shape of the non-dominated set. In practical situations, the preliminary preferences of the DM are often non-realistic and his/her expectations usually far exceed attainable ranges of objectives. The DM is 'learning' about the problem during the interactive process. Wavering, incoherence and changes of DM's preferences are typical to the process. So, the more the DM learns about the non-dominated set at each iteration, the fewer steps are necessary to find a final solution and the stronger is the conviction of the DM that he/she has found the best compromise. Another drawback of methods from this class is that no information about a neighbourhood of the current point is presented to the DM. So, the DM can miss a possibility of improving the score on one objective at a very small expense of other objectives.

There is also a class of interactive procedures that present to the DM samples of non-dominated points at each iteration. To this class belong such methods like: Zionts-Wallenius method (Zionts and Wallenius, 1976), Jacquet-Lagrè ze, Meziani and Słowiński method (Jacquet-Lagrè ze et al., 1987), as well as the Reference Direction Approach (Narula et al., 1992) which is an extension of the VIG method (Korhonen, 1987) for the non-linear case, and Computer Graphics-Based method (Korhonen et al., 1992) which is another extension of VIG. At decision phases of such methods, the DM is usually expected to evaluate the presented solutions and specify which one is the best or rank all the solutions in the sample. Authors of these methods make an assumption that evaluation of a finite sample of non-dominated points is relatively easy for the DM.

However, it follows from practical experience and theoretical works made in the field of MCDA that evaluation of an even small finite set of alternatives can be difficult to the DM. It is rather illusory to expect from the DM an explicit and complete evaluation of the alternatives if, for example, some of them are incomparable. Instead, he/she gives some preferential information upon which a global preference model can be built.

The above mentioned procedures can fail if the DM refuses to accept a substitution between objectives. Such a situation arises when objectives are in strong conflict. In this case, the DM may be simply unable to compare alternatives that are significantly different. Another type of difficulties may appear if the values of objective functions calculated for a feasible solution are uncertain for some reasons. In this case, small differences in the values of the objective functions are meaningless for the DM and alternatives that are not different enough might be considered indifferent.

It is usually assumed that one of the four following situations can appear while comparing two alternatives a and b (Vincke, 1990):

a P b i.e. *a* is preferred to *b*,

b P a i.e. *b* is preferred to *a*,

a I b i.e. *a* and *b* are indifferent,

a ? b i.e. *a* and *b* are incomparable.

The *preference P*, *indifference I* and *incomparability* ? relations are the sets of ordered pairs (*a*, *b*) such that *a P b*, *a I b*, *a ? b*, respectively. The relations are not assumed to be transitive.

However, in order to handle situations where the DM is not able or unwilling to make distinctions between *a P b*, *a I b* and *a ? b*, the use is recommended of a grouped relation *S* called an *outranking relation* (Roy, 1985): *a S b* means that *a* is at least as good as *b*; $a \not S b$ and $b \not S a$ mean that *a* and *b* are incomparable.

In order that each particular step of an interactive procedure makes an improvement in the search for the best compromise solution, the sample of points presented to the DM for an examination should meet some requirements. Specifically, the points in the sample should not be indifferent nor incomparable. Otherwise, difficulties in evaluation of the sample can yield additional incoherence in the preferential information supplied by the DM. Moreover, in such a case, the DM can stop the interactive procedure being not able to find a better proposal among the presented points even if the current point is far from the best compromise.

The procedure presented in this paper tries to overcome the drawbacks of the above mentioned interactive procedures. Specifically,

- it uses an outranking relation as a local preference model built in a neighbourhood of a current point,
- the neighbourhood of the current point is composed of non-dominated points that outrank this point, so the neighbourhood includes points that are sufficiently different but comparable; the points from outside the neighbourhood are either incomparable or outranked by the current point,
- the sample of non-dominated points presented to the DM in each decision phase comes from the neighbourhood of the current point,
- the outranking relation used to define the interesting sub-region of the non-dominated set is based on a relatively weak preferential information of an inter- and intra-criteria type,
- the scanning of the non-dominated set is organized such that the sub-region moves in result of either a change of the DM's reference point or a shift of the current point within a neighbourhood of this point.

The last point submits some analogy with projecting of a focused beam of light from a spotlight in the reference point onto the non-dominated set. For this reason, we call our procedure the Light Beam Search or, shortly, LBS (see figure 1).

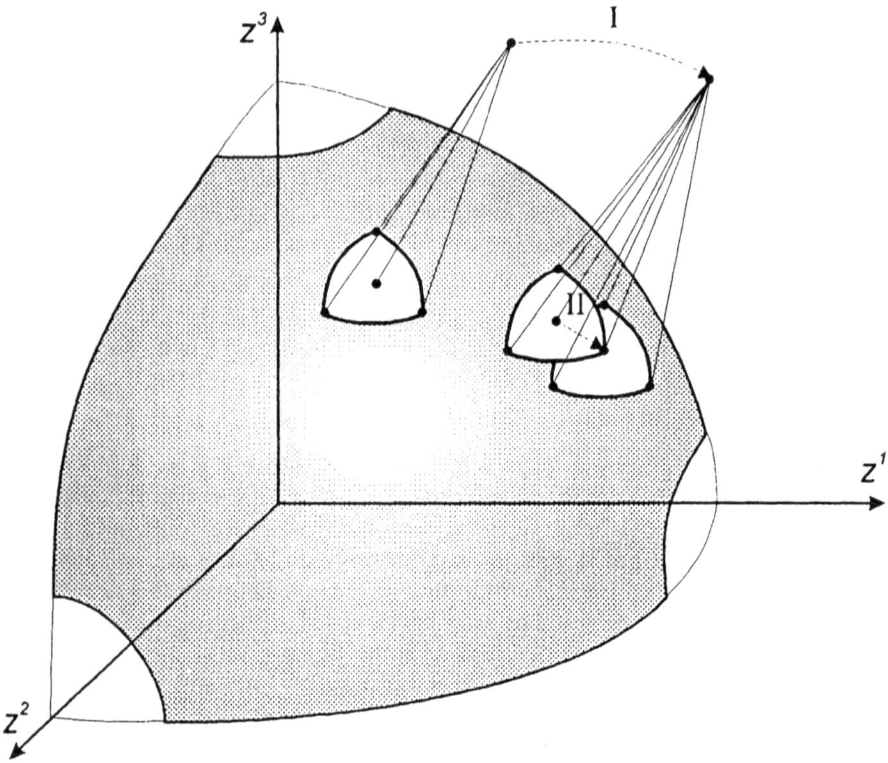

I - Change of the reference point

II - Shift of the middle point

- The non-dominated set

- Highlighted non-dominated neighbourhood of the middle point

Figure 1. The Light Beam Search over a non-dominated set

The paper is organized in the following way. After a formal statement of the problem and basic definitions, the main idea of the procedure will be presented. Then, the general scheme of the procedure will be outlined. Some characteristic points of the procedure will be described more precisely in the following section. In the next section, an illustrative example will be solved with the LBS procedure. Finally, main features of the procedure will be summarized and some possible directions of further researches will be pointed out.

2. Problem statement and basic defintions

The general multiple-objective programming problem is formulated as

$$\max\{f_1(\mathbf{x}) = z_1\}$$
$$\cdots\cdots\cdots\cdots \qquad \text{(P1)}$$
$$\max\{f_k(\mathbf{x}) = z_k\}$$

s.t.

$$\mathbf{x} \in D$$

where $\mathbf{x} = [x_1,...,x_n]$ is a vector of decision variables, functions f_j are continuous and differentiable and condition $\mathbf{x} \in D$ can be stated using continuous and differentiable constraints.

Problem (P1) can also be stated more succinctly as

$$\max\{\mathbf{z}\} \qquad \text{(P2)}$$

s.t.

$$\mathbf{z} \in Z$$

where $\mathbf{z} = [z_1,...,z_k]$ is a vector of objective functions $z_j = f_j(\mathbf{x})$ and Z is an image of set D in the objective space.

Point $\mathbf{z}' \in Z$ is *non-dominated* if there is no $\mathbf{z} \in Z$ such that $z_j \geq z_j' \; \forall j$, and $z_i > z_i'$ for at least one i. Point $\mathbf{z}' \in Z$ is *weakly non-dominated* if there is no $\mathbf{z} \in Z$ *such that* $z_j > z_j' \; \forall j$. The set of all non-dominated points is the *non-dominated set*. For other definitions concerning non-dominance and efficiency, see e.g. Wierzbicki (1986).

The point $\mathbf{z}^*$ composed of the best attainable objective function values is called the *ideal point*

$$z_j^* = \max\ \{f_j(\mathbf{x}) \mid \mathbf{x} \in D\} \qquad j = 1,...,k.$$

Another useful definition is that of the *achievement scalarizing function* in the objective space:

$$s(\mathbf{z},\Lambda,\rho) = \max_j\left\{\lambda_j\left(\varepsilon_j + z_j^0 - z_j\right)\right\} + \rho\sum_{j=1}^{J}\left(z_j^0 - z_j\right) \qquad (1)$$

where $\mathbf{z}^0$ is a reference point, $\varepsilon_j > 0$ is moderately small, $\Lambda = [\lambda_1,...,\lambda_k]$ is a weighting vector, $\lambda_i \geq 0$, $\sum_{j=1}^{k}\lambda_j = 1$ and ρ is a sufficiently small positive number.

3. Main idea of the Light Beam Search procedure

The LBS procedure falls into the category of interactive procedures with generation of finite samples of non-dominated points at each computation phase. A sample is composed of a current point, called the *middle point*, obtained at a previous iteration, and a number of non-dominated points from its neighbourhood. In order to define the neighbourhood the sample represents, we use an outranking relation S as a local preference model. Precisely, for a current middle point, the sub-region is defined as a set of non-dominated points that are not worse than the middle point, i.e. outrank the middle point. The sub-region is called the *outranking neighbourhood* of the middle point. The sample is composed of points that are obtained by independent optimization of particular objectives in the outranking neighbourhood, called the *characteristic neighbours* of the middle point. Moreover, the DM is able to scan more precisely the inner of the neighbourhood through the objective function trajectories between any two characteristic neighbours or between a characteristic neighbour and the middle point. Other methods for exploration of the neighbourhood can also be used.

The formal expression of the conditions that must be satisfied to validate the assertion $a\,S\,b$ can be influenced by many factors. In the presented procedure, following the approaches proposed in various versions of ELECTRE methods (Roy, 1990), the following factors will be taken into account:

- the discrimination power of the DM's preferences with respect to particular objectives that will be modelled with *indifference* and *preference thresholds* (i.e. the intra-criteria information),
- the inter-criteria information specified in the form of the *veto thresholds*.

Similarly to ELECTRE IV, we will not use the inter-criteria information in the form of importance coefficients which might be too difficult to define; we assume, however, that one objective is not more important that all the others together. Let us notice that the ratio of veto and preference thresholds of a criterion is related with its importance; the lower is the ratio the greater is the importance (Roy, 1980).

In the traditional preference modelling it is assumed that every difference on a single objective z_j is significant to the DM. However, in practice, there exists an interval in which the DM does not feel any difference between two elements or refuses to accept a preference for one of the alternatives. This fact was already pointed out by Poincaré (1935 p.69), but it was Luce (1956) who introduced this fundamental remark to preference modelling. This situation can be modelled with the indifference threshold q_j given by the DM.

Moreover, experience shows that, usually, there is no precise value giving the limit between the indifference and preference but there exists an intermediary region where the DM hesitates between the indifference and preference or gives different answers, depending on the way he/she is questioned. This remark has led to the introduction of the preference threshold p_j. In general, the indifference and preference thresholds are functions of z_j; moreover:

$$p_j(z_j) \geq q_j(z_j) \geq 0$$

The indifference and preference thresholds allow to distinguish between the three following preference relations with respect to z_j for any ordered pair (a, b) of alternatives:

$a\, I_j\, b$ i.e. a and b are *equivalent* $\Leftrightarrow$ $-q_j\left(z_j^a\right) \le z_j^a - z_j^b \le q_j\left(z_j^b\right)$,

$a\, Q_j\, b$ i.e. a is *weakly preferred* to b $\Leftrightarrow$ $q_j\left(z_j^a\right) < z_j^a - z_j^b < p_j\left(z_j^b\right)$,

$a\, P_j\, b$ i.e. a is *strictly preferred* to b $\Leftrightarrow$ $p_j\left(z_j^a\right) \le z_j^a - z_j^b$,

The veto threshold v_j allows to take into account the possible difficulties of comparing the relative value of two alternatives when one is significantly better than the other on a subset of objectives, but much worse on at least one other objective. In general, the veto threshold is also a function of z_j.

The outranking relation has already been used as a preference model in the Cone Contraction Method with Visual Interaction for Multiple-Objective Non-Linear Programmes (Jaszkiewicz and Słowiński, 1992). In that method, however, it is used as global preference model. The construction of an outranking relation follows the methodology proposed for the ELECTRE III method (Roy, 1978) and the relation is built on a representative sample of non-dominated points. As the indifference, preference and veto thresholds, in general, depend on z_j, the DM should specify these thresholds in the form of mathematical functions, $q_j(z_j)$, $p_j(z_j)$ and $v_j(z_j)$. If the functions are complicated, it is practically impossible for the DM to specify them explicitly. In the Light Beam Search procedure the outranking relation is used as a local preference model in the neighbourhood of a middle point, so a single value of each threshold is sufficient for a given middle point. Of course, the DM can update the values of the thresholds for every new middle point.

The outranking relation has also been used as a local preference model in the method proposed by Lotfi et al. (1992). However, their method is purposed for multiple objective analysis of problems with finite set of alternatives only. In this case, the whole neighbourhood can be generated and presented to the DM. Moreover, as the authors do not use any additional preferential information, the definition of the neighbourhood seems somewhat arbitrary.

4. General scheme of the interactive procedure

The following is a general scheme of the proposed procedure presented in a Pascal-like form:

> Fix the points of the best and the worst values of objectives; make the former one the first reference point;
> **if** *the DM wants to specify the first reference point* **then**
>   Ask the DM to specify the aspiration levels on particular objectives;
> Find a starting middle point;

Ask the DM to specify the preferential information of inter- and intra-criteria type;

repeat

Present the middle point to the DM;

Calculate the characteristic neighbours of the middle point and present them to the DM;

Allow the DM to scan the inner area of the current neighbourhood;

if *the DM wants to store the middle point* **then**

Add it to the set of stored points;

case

The DM wants to define a new reference point:

Ask the DM to specify the aspiration levels on particular objectives;

Project the reference point onto the non-dominated set;

The DM wants a point from the neighbourhood to be the new middle point:

Ask the DM to select the new middle point;

The DM wants to return to one of the stored points:

Use the selected stored point as a new middle point;

The DM wants to update the preferential information:

Ask the DM to specify the new preferential information;

end

until *the DM feels satisfied with a point found during the interactive process or concludes that no compromise point exists for the present constraints;*

5. Detailed description of the particular steps

The procedure starts by asking the DM to specify (subjective) best and worst values of objectives, z_j^*, z_{j*} ($j = 1,..,k$), respectively. If he/she is unable to do so, the best values are fixed at individual maxima of particular objectives (ideal point) and the worst values are set equal to minimal values of objectives at the points corresponding to the individual maxima. The point of the best values $\mathbf{z}^*$ becomes the first reference point, $\mathbf{z}^o$.

Then, the DM is asked to give the preferential information for each objective, i.e. the indifference and, optionally, the preference and veto threshold. At this stage the DM should decide if he/she wants to specify the preference and/or veto thresholds, however, he/she is able to change these settings at every step of the procedure.

In the next step, the starting middle point $\mathbf{z}^c$ is computed. The point is obtained by projecting point $\mathbf{z}^*$ of the best values of objectives onto the non-dominated set in the direction defined by point $\mathbf{z}^*$ and point $\mathbf{z}_*$ of the worst values of objectives. The achievement scalarizing function (1) is used to this end.

Then, the *characteristic neighbours* of the middle point are computed. The characteristic neighbour, with respect to objective z_j is a point $\mathbf{z}^j$ from the outranking neighbourhood of point $\mathbf{z}^c$ that maximizes the distance from $\mathbf{z}^c$ in direction of the greatest locally feasible improvement of objective z_j ($j = 1,..,k$). An *attainable characteristic neighbour* $\mathbf{z}'^j$ is a point obtained in result of a projection of point $\mathbf{z}^j$ onto the non-dominated set ($j = 1,..,k$).

In order to test if a point $\mathbf{z}$ outranks the middle point, first, the following numbers are calculated:

$m_s(\mathbf{z}, \mathbf{z}^c)$ - the number of the objectives for which point $\mathbf{z}$ is indifferent, or weakly or strictly preferred, to $\mathbf{z}^c$,

$m_q(\mathbf{z}^c, \mathbf{z})$ - the number of the objectives for which point $\mathbf{z}^c$ is weakly preferred to $\mathbf{z}$,

$m_p(\mathbf{z}^c, \mathbf{z})$ - the number of the objectives for which point $\mathbf{z}^c$ is strictly preferred to $\mathbf{z}$,

$m_v(\mathbf{z}^c, \mathbf{z})$ - the number of the objectives being in a strong opposition to the assertion $\mathbf{z}\,S\,\mathbf{z}^c$, i.e. card $\{j\colon z_j^c - v_j \geq z_j, j = 1,..,k\}$.

The construction of the outranking relation depends on the type of preferential information supplied by the DM. If the DM has specified all of the thresholds, we propose to use the following definition of the outranking relation:

$$\mathbf{z}\,S^a\,\mathbf{z}^c \Leftrightarrow \begin{cases} m_v(\mathbf{z}^c,\mathbf{z}) = 0 \text{ and} \\ m_p(\mathbf{z}^c,\mathbf{z}) \leq 1 \text{ and} \\ m_q(\mathbf{z}^c,\mathbf{z}) + m_p(\mathbf{z}^c,\mathbf{z}) \leq m_s(\mathbf{z},\mathbf{z}^c) \end{cases}$$

If the DM has decided not to specify the veto thresholds, we should not assume that for every objective he/she is ready to accept any worsening of the objective even if a subset of other objectives is significantly improved, i.e. we should not assume that the veto threshold does not exist. Such a situation indicates that at the particular stage of the interactive process, the DM is not able or unwilling to specify the value of this threshold explicitly. In this case, we propose to use the following definition of the outranking relation:

$$\mathbf{z}\,S^b\,\mathbf{z}^c \Leftrightarrow \begin{cases} m_p(\mathbf{z}^c,\mathbf{z}) = 0 \text{ and} \\ m_q(\mathbf{z}^c,\mathbf{z}) \leq m_s(\mathbf{z},\mathbf{z}^c) \end{cases}$$

Let us observe that $S^b \subseteq S^a$ and that $S^b = S^a$ if $v_j = p_j\ \forall j$. If, for an objective z_j, the DM is ready to accept the worsening of its value greater that p_j, some points that outrank the middle point can be left outside the outranking neighbourhood. However, the neighbourhood will be still composed of points that are comparable with the middle point.

In a similar way we can analyze the situation when the DM has decided not to specify the preference threshold. In this case we should not assume that the DM feels no difference between the weak and strict preferences. Such a situation indicates that the DM is not able or unwilling to specify explicitly the value allowing to distinguish between the two relations. In this case, we propose to use the following definition of the outranking relation:

$$\mathbf{z}\,S^c\,\mathbf{z}^c \Leftrightarrow \begin{cases} m_v(\mathbf{z}^c,\mathbf{z}) = 0 \text{ and} \\ m_p(\mathbf{z}^c,\mathbf{z}) \le 1 \end{cases}$$

Let us observe that $S^c \subseteq S^a$ and that $S^c = S^a$ if $p_j = q_j\ \forall j$.

Finally, if the DM has decided to specify the indifference thresholds only, we propose to use the following definition of the outranking relation:

$$\mathbf{z}\,S^d\,\mathbf{z}^c \Leftrightarrow m_p(\mathbf{z}^c,\mathbf{z}) = 0$$

Let us observe that $S^d \subseteq S^b \subseteq S^a$ and $S^d \subseteq S^c \subseteq S^a$, moreover, $S^d = S^a$ if $v_j = p_j = q_j\ \forall j$.

In order to find the k characteristic neighbours, gradients of particular objectives are projected onto linear approximation of constraints which are active in point $\mathbf{z}^c$ (cf. gradient projection methods for non-linear optimization, Rosen, 1960). Let h be the number of active constraints in point $\mathbf{z}^c$. The linear constraints can be presented in a matrix form

$$\mathbf{Ax} = \mathbf{b}$$

where $z_j^c = f_j(\mathbf{x}^c), j = 1,...,k,\ a_{ti} = \partial c_t / \partial x_i$ are elements of matrix $\mathbf{A}$, t is an index of an active constraint, $t = 1,..,h$; $i = 1,...,n$. Next, the projection matrix $\mathbf{P}$ is calculated

$$\mathbf{P} = \mathbf{I} - \mathbf{A}^T(\mathbf{AA}^T)^{-1}\mathbf{A}$$

Matrix $\mathbf{P}$ and gradients of particular objectives $\nabla_x f_j$ are used to obtain directions $\Delta\mathbf{x}^j$ in the space of variables

$$\Delta\mathbf{x}^j = \mathbf{P}\nabla_x f_j, \quad j = 1,...,k$$

$\Delta\mathbf{x}^j$ is the feasible direction of the greatest improvement of objective $z_j = f_j(\mathbf{x})$. Directions $\Delta\mathbf{x}^j$ are used in turn to define corresponding directions $\mathbf{V}^j$ in the objective space

$$\mathbf{V}^j = [\sum_{i=1}^{n}(\Delta x_i^j \partial f_1 / \partial x_i),...,\sum_{i=1}^{n}(\Delta x_i^j \partial f_k / \partial x_i)]$$

Then, the following mathematical programming problem is solved in order to maximize objective z_j in direction $\mathbf{V}^j$ ($j = 1,...,k$)

$$\begin{aligned} & \max\ \alpha \\ \text{s.t.}\quad & \mathbf{z}^j\,S\,\mathbf{z}^c, \qquad \mathbf{z}^j = \mathbf{z}^c + \alpha\,\mathbf{V}^j \\ & \alpha \ge 0 \end{aligned} \qquad \text{(P3)}$$

Problem (P3) is a small mathematical programming problem with one variable only. Points, $\mathbf{z}^j$ ($j = 1,...,k$) following from solving of k problems (P3) give characteristic neighbours (see figure 2).

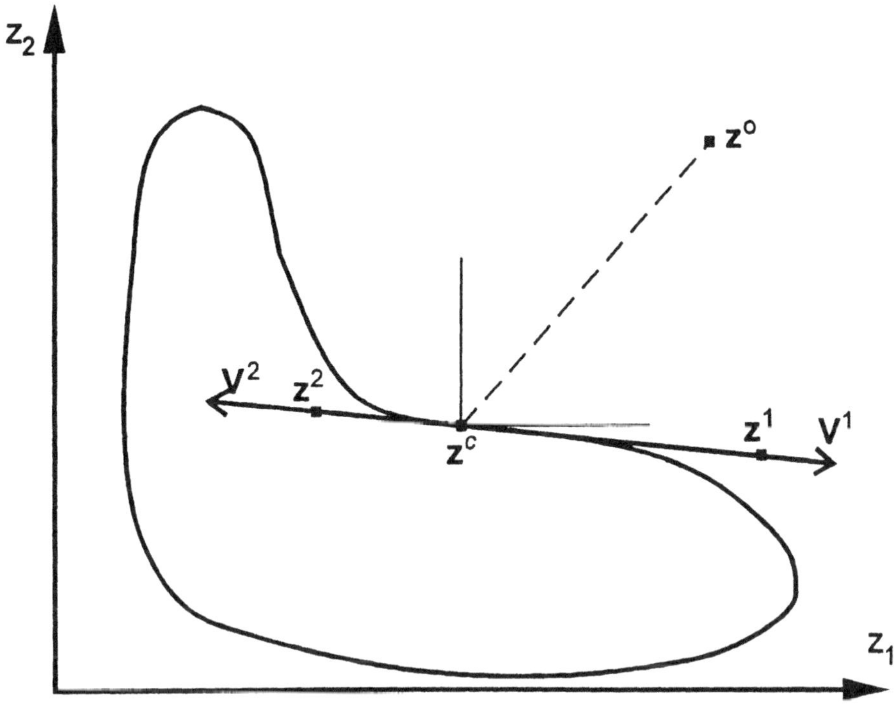

Figure 2. Characteristic neighbours found using a gradient projection onto a linear approximation of active constraints in z^c

Attainable characteristic neighbours are obtained in result of projection of points $\mathbf{z}^j$ ($j = 1,..,k$) onto the non-dominated set in direction connecting $\mathbf{z}^j$ with point $\mathbf{z}_*$ (see figure 3).

In the decision phase, the middle point and its characteristic neighbours are presented to the DM. Both numerical and graphical forms of presentation should be used to help the DM in evaluating large amounts of information. Moreover, the DM is able to scan more precisely the region between any two characteristic neighbours or between a characteristic neighbour and the middle point. For this purpose, the line segment connecting the points in the objective space is projected onto the non-dominated set. The obtained subset of the non-dominated points is called the *profile* of the neighbourhood. As in non-linear case getting a continuous profile is practically impossible, we choose a finite numbers of points lying on the line segment and project them onto the non-dominated set (see figure 4). The points resulting from the projection are then presented to the DM. A similar technique of scanning a sub-region of the non-dominated set has been used in Jaszkiewicz and Słowiński (1992). Some other techniques of local characterisation of the non-dominated set can also be used at this step.

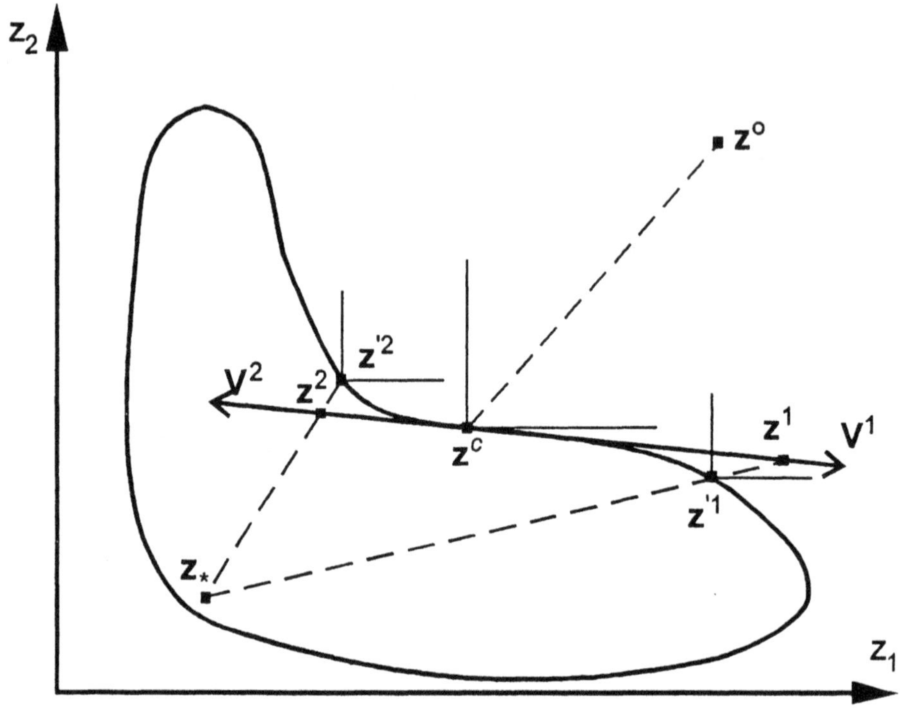

Figure 3. Attainable characteristic neighbours found by projecting points z^1 and z^2 onto the non-dominated set

The procedure stops if one of the presented points is satisfactory to the DM on all objectives. Otherwise, he/she can continue the scanning using two degrees of freedom. The first degree consists in modifying the aspiration levels, i.e. the reference point. The new reference point is then projected onto the non-dominated set in order to find the new middle point. The second degree of freedom consists in selecting one of the points from the neighbourhood to be the new middle point for the same reference point. Then a new outranking neighbourhood is generated (see figure 1).

Before continuing the scanning, the DM can store the current middle point. He/she is allowed to restore any of the stored points at any time.

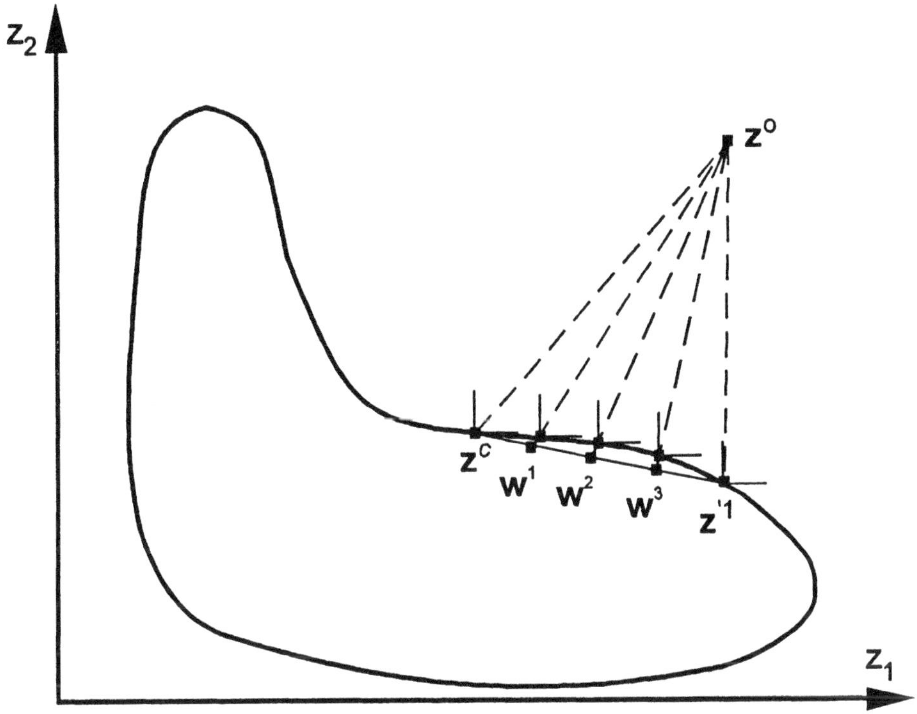

Figure 4. Finding an approximation of a profile of the non-dominated set

Finally, the DM is able to modify the preferential information given for each objective, i.e. the indifference, preference an veto thresholds. He/she can also change the type of the outranking relation. It influences the construction of the outranking relation and the size of the new outranking neighbourhood.

6. Illustrative example

To illustrate the work of the Light Beam Search procedure the following example will be solved

$$\max\left\{f_1 = x_1^2 - 0.1x_2 - 0.1x_3\right\}$$

$$\max\left\{f_1 = x_2^2 - 0.1x_1 - 0.1x_3\right\}$$

$$\max\left\{f_1 = x_3^2 - 0.1x_1 - 0.1x_2\right\}$$

s.t.

$$x_1^3 + x_2^3 + x_3^3 \leq 24$$

$$x_1 \geq 0$$

$$x_2 \geq 0$$

$$x_3 \geq 0$$

The procedure starts by fixing the points of the best and the worst values of objectives, $\mathbf{z}^*$ and $\mathbf{z}_*$, respectively. Assume that the DM is not willing to specify the points and they are fixed automatically as ideal and nadir points

$$\mathbf{z}^* = [8.32, 8.32, 8.32]$$

$$\mathbf{z}_* = [-0.288, -0.288, -0.288]$$

The procedure finds the starting middle point

$$\mathbf{z}^c = [3.6, 3.6, 3.6]$$

Then the DM is asked to decide what kind of the preferential information he wants to specify. The DM decides to specify all the thresholds and gives the following values

Objective	q_j	p_j	v_j
f_1	0.1	0.5	1.5
f_2	0.2	0.8	3
f_3	0.4	0.9	6

An outranking neighbourhood is constructed around the middle point. Attainable characteristic neighbours coming from the neighbourhood are calculated

$$\mathbf{z}^1 = [4.380, 3.185, 3.185]$$

$$\mathbf{z}^2 = [3.185, 4.380, 3.185]$$

$$\mathbf{z}^3 = [3.396, 3.396, 3.995]$$

The points are presented to the DM. The DM has an idea about desired values of objectives and decides to specify a reference point. He/she gives the following point

$$\mathbf{z}^o = [7.2, 5.8, 2.8]$$

The point given by the DM is non-attainable. The procedure projects it onto the non-dominated set. The new middle point

$$\mathbf{z}^c = [5.338, 3.938, 0.938]$$

is a result of the projection. The attainable characteristic neighbours found for the new middle points are the following

$$\mathbf{z}^1 = [7.106, 1.797, 0.508]$$

$$\mathbf{z}^2 = [3.742,\ 5.636,\ 0.697]$$

$$\mathbf{z}^3 = [5.047,\ 3.720,\ 1.722]$$

The DM feels that at this point the values of some thresholds are not appropriate and decides to change the preferential information. This time he/she gives the following values of the thresholds

Objective	q_j	p_j	v_j
f_1	0.05	0.3	0.8
f_2	0.2	0.5	3
f_3	0.3	0.7	6

With the new preferential information the procedure constructs a new outranking neighbourhood and calculates new attainable characteristic neighbours

$$\mathbf{z}^1 = [6.724,\ 2.336,\ 0.619]$$

$$\mathbf{z}^2 = [4.505,\ 4.884,\ 0.815]$$

$$\mathbf{z}^3 = [5.047,\ 3.720,\ 1.722]$$

The DM thinks that point $\mathbf{z}^3$ is better than the middle points and selects it to be the new middle point. New attainable characteristic neighbours are found and presented to the DM

$$\mathbf{z}^1 = [5.884,\ 2.937,\ 1.411]$$

$$\mathbf{z}^2 = [4.210,\ 4.755,\ 1.485]$$

$$\mathbf{z}^3 = [4.749,\ 3.508,\ 2.393]$$

The DM decides to scan more precisely the profile between the middle point $\mathbf{z}^c$ and its characteristic neighbour $\mathbf{z}^3$. The profile is constructed by projection of the line segment that connects point $\mathbf{z}^c$ with $\mathbf{z}^3$. The sample of the non-dominated points from the profile is presented to the DM

$$\mathbf{w}^1 = [5.006,\ 3.691,\ 1.819]$$

$$\mathbf{w}^2 = [4.964,\ 3.662,\ 1.916]$$

$$\mathbf{w}^3 = [4.922,\ 3.632,\ 2.013]$$

$$\mathbf{w}^4 = [4.879,\ 3.602,\ 2.109]$$

$$\mathbf{w}^5 = [4.836,\ 3.571,\ 2.204]$$

$$\mathbf{w}^6 = [4.793,\ 3.540,\ 2.299]$$

The DM feels that point $\mathbf{w}^2$ is satisfactory on all objectives. Thus it yields the best compromise. In the space of variables the point correspond to the following solution

$$x_1 = 2.306$$

$$x_2 = 2.011$$

$$x_3 = 1.532$$

At this point the first constraint is active.

7. Summary and conclusions

A new interactive procedure for multiple-objective analysis of linear and non-linear programs has been presented. At the decision phase of the procedure a sample of non-dominated points is presented to the DM. The sample comes from an outranking neighbourhood of a given middle point. The DM has two degrees of freedom to move from one neighbourhood to another: by changing his/her aspiration levels or by selecting one of the points from the neighbourhood to be the new middle point. The procedure is called the Light Beam Search. Its microcomputer implementation working under MS-Windows environment is available from the authors upon request. The class of non-linear problems that can be solved with the implementation depends on the type of a non-linear solver used; non-convex problems require specialized solvers giving a relatively good chance of obtaining the global optimum.

Although, the presented procedure has been developed for analysis of continuous mathematical programming problems, it can be extended for discrete programming problems and for problems with finite sets of alternatives. In these cases, it could be possible to present all points from the outranking neighbourhood to the DM. Such techniques like filtering and clustering can be used to reduce the number of points presented at the decision phase if the number of points belonging to the outranking neighbourhood would be too large.

In general, a considerable computational effort is required to find the characteristic neighbours of the middle point in an exact way. A parallel computing can be used to accelerate calculations at the computation phase. Moreover, even on a sequential computer, it is possible to minimize the durations of particular steps by overlapping the computation and decision phases. To this end, the middle point should be presented to the DM as soon as it is found. The DM should also be able to use various forms of presentation of this point and to change it before all the characteristic neighbours are found. Furthermore, every characteristic neighbour should be presented to the DM as soon as it is calculated. As a dialogue with the user does not cause a great burden of the CPU it is possible to perform the calculations in the background.

Acknowledgment

This work has been supported by KBN grant no. 43-232/94/DS.

References

Benayoun R., de Montgolfier J., Tergny J. and Larichev O. (1971). Linear programming with multiple objective functions: step method (STEM). *Operational Research Quarterly*, **24**, 65-77.

Jacquet-Lagrè ze E., Meziani R. and Slowiński R. (1987). MOLP with an interactive assessment of a piecewise-linear utility function. *Eur. J. Oper. Res.*, **31**, 350-357.

Jaszkiewicz A. and Słowiński R. (1992). Cone Contraction Method with Visual Interaction for Multiple-Objective Non-Linear Programmes. *Journal of Multi-Criteria Decision Analysis*, **1**, 29-46.

Korhonen P. (1987). VIG - a visual interactive support system for multiple criteria decision making. *Belg. J. Ops Res. Statist. Comput. Sci.*, **27**, 3-15.

Korhonen P. and Wallenius J. (1988). A Pareto race. *Naval Research Logistics*, **35**, 615-623.

Korhonen P., Wallenius J. and Zionts S. (1992). A Computer Graphics-Based Decision Support System for Multiple Objective Linear Programming. *Eur. J. Oper. Res.*, **60**, 280-286.

Lee S.M. and Shim J.P. (1986). Interactive goal programming on the microcomputer to establish priorities for small business. *Journal of the Operational Research Society*, **37**, 571-577.

Lotfi V., Stewart T.J. and Zionts S. (1992). An aspiration-level interactive model for multiple criteria decision making. *Comput. Oper. Res.*, in press.

Luce D. (1956). Semiorders and a theory of utility discrimination. *Econometrica*, **24**, 178-191.

Narula S.C., Kirilov L. and Vassilev V. (1992). Reference direction approach for solving multiple objective nonlinear programming problems. *Proceedings of the Tenth International Conference on Multiple Criteria Decision Making*, Taipei, **2**, 355-362.

Poincaré H. (1935). *La valeur de la Science*, Flammarion, Paris.

Rosen J.B. (1960). The gradient projection method for nonlinear programming. Part I: Linear constraints. *SIAM J. Appl. Math.*, **8**.

Roy B. (1978). ELECTRE III - un algorithme de classement fondé sur une représentation floue des préférences en présence de critè res multiples. *Cahiers du CERO*, **20**, 3-24.

Roy B. (1985). *Méthodologie Multicritè re d'Aide à la Décision*, Economica, Paris, 1985.

Roy B. (1990). The outranking approach and the foundations of ELECTRE methods. In Bana e Costa C.A. (ed.) *Readings in Multiple Criteria Decision Aid*, Springer-Verlag, Berlin, 155-183.

Vincke P. (1990). Basic concepts of preference modelling. In Bana e Costa C.A. (ed.) *Readings in Multiple Criteria Decision Aid*, Springer-Verlag, Berlin, 101-118.

Wierzbicki A.P. (1980). The use of reference objective in Multiobjective Optimization. In Fandel G. and Gal T. (eds.) *Multiple Criteria Decision Making, Theory and Application*, Springer-Verlag, Berlin, 468-486.

Wierzbicki A.P. (1986). On the completeness and constructiveness of parametric characterizations to vector optimization problems. *OR Spectrum*, **8**, 73-87.

Zionts S. and Wallenius J. (1976). An interactive programming method for solving the multiple criteria problem. *Management Science*, **22**, 652-663.

ASPIRATION LEVEL APPROACH TO INTERACTIVE MULTI-OBJECTIVE PROGRAMMING AND ITS APPLICATIONS

Hirotaka Nakayama
Department of Applied Mathematics
Konan University
8-9-1 Okamoto, Higashinada, Kobe 658, JAPAN
e-mail: nakayama@edu2.math.konan-u.ac.jp

Abstract

Several kinds of techniques for multiple criteria decision making have been developed for the last few decades. Above all, the aspiration level approach to multi-objective programming problems is widely recognized to be effective in many practical fields. As one of the aspiration level approach, the author developed the satisficing trade-off method. In addition, he has been applying the method to several kinds of practical problems for these ten years. Some of them were already performed in real life. Typical examples such as feed formulation for live stock, erection management of cable stayed bridge and bond portfolio will be included in this paper.

Keywords: multi-objective programming, applications of multi-objective programming, satisficing trade-off method, aspiration level approach

1 Introduction

Multi-objective programming problems are formulated as follows:

P.M. Pardalos et al. (eds.), Advances in Multicriteria Analysis, 147–174.

$$\text{(MOP)} \qquad \text{Minimize} \quad f(x) \equiv (f_1(x),\ f_2(x), \ldots,\ f_r(x))$$
$$\text{over} \quad x \in X.$$

The constraint set X may be given by

$$g_j(x) \leq 0, \qquad j = 1, \ldots, m,$$

and/or a subset of R^n itself. For the problem (MOP), we define Pareto solutions as follows:

Definition 1.1 A solution $\hat{x}$ is said *Pareto optimal,* if there is no better solution $x \in X$ other than $\hat{x}$, namely, if

$$f(x) \not\leq f(\hat{x}) \qquad \text{for any} \quad x \neq \hat{x} \in X.$$

In general, there may be many Pareto solutions. The final decision is made among them taking the total balance over all criteria into account. This is a problem of value judgment of decision maker (in abbreviation, DM). The totally balancing over criteria is usually called *trade-off.* It should be noted that there are very many criteria, say, over one hundred in some practical problems such as erection management of cable stayed bridge, and camera lens design. Therefore, it is very important to develop effective methods for helping DM to trade-off easily even in problems with very many criteria.

Interactive multi-objective programming search a solution in an interactive way with DM while eliciting information on his/her value judgment. Along this line, several methods have been developed remarkably for the last about fifteen years: Among them, the aspiration level approach is now recognized very effective in practice, because

(i) it does not require any consistency of DM's judgment,

(ii) aspiration levels reflect the wish of DM very well,

and

(iii) aspiration levels play the role of probe better than the weight for objective functions.

In the following, we will discuss the difficulty in weighting method which is commonly used in the traditional goal programming.

2 Why is the Weighting Method Ineffective?

In multi-objective programming problems, the final decision is made on the basis of the value judgment of DM. Hence it is important how we elicit the value judgment of DM. In many practical cases, the vector objective function is scalarized in such a manner that the value judgment of DM can be incorporated.

The most well known scalarization technique is the linearly weighted sum:

$$\sum_{i=1}^{r} w_i f_i(x). \tag{2.1}$$

The value judgment of DM is reflected by the weight. Although this type of scalarization is widely used in many practical problems, there is serious drawbacks in it. Namely, it can not provide a solution among sunken parts of Pareto surface due to "*duality gap*" for nonconvex cases. Even for convex cases, for example, in linear cases, even if we want to get a point in the middle of line segment between two vertices, we merely get a vertex of Pareto surface, as long as the well known simplex method is used. This implies that depending on the structure of problem, the linearly weighted sum can not necessarily provide a solution as DM desires.

In the traditional goal programming (Charnes-Cooper, 1961), some kind of metric function from the goal f^* is used as the one representing the preference of DM. For example, the following is well known:

$$\left(\sum_{i=1}^{r} w_i |f_i(x) - f_i^*|^p\right)^{1/p} \tag{2.2}$$

The preference of DM is reflected by the weight w_i, the value of p, and the value of the goal f_i^*. If the value of p is chosen appropriately, a Pareto solution among a sunken part of Pareto surface can be obtained by minimizing the function (2.2). However, it is usually difficult to pre-determine appropriate values of them. Moreover, the solution minimizing (2.2) can not be better than the goal $\boldsymbol{f}^*$, even though the goal is underestimated.

In addition, one of the most serious drawbacks in the goal programming is that people tend to misunderstand that a desirable solution can be obtained by adjusting the weight. It should be noted that there is no positive correlation between the weight w_i and the value $f(\hat{x})$ corresponding to the resulting solution $\hat{x}$ as will be seen in the following example.

Example 2.1 Let $y_1 = f_1(x)$, $y_2 = f_2(x)$ and $y_3 = f_3(x)$, and let the feasible region in the objective space be given by

$$\{(y_1, y_2, y_3) |\ (y_1 - 1))^2 + (y_2 - 1)^2 + (y_3 - 1)^2 \leq 1\}.$$

Suppose that the goal is $(y_1^*,\ y_2^*,\ y_3^*) = (0,\ 0,\ 0)$.The solution minimizing the metric function (2.2) with $p = 1$ and $w_1 = w_2 = w_3 = 1$ is $(y_1,\ y_2,\ y_3) = (1 - 1/\sqrt{3},\ 1 - 1/\sqrt{3},\ 1 - 1/\sqrt{3})$. Now suppose that DM wants to decrease the value of f_1 a lot more and that of f_2 a little more, and hence modify the weight into $w_1' = 10,\ w_2' = 2,\ w_3' = 1$. The solution associated with the new weight is $(1 - 10/\sqrt{105},\ 1 - 2/\sqrt{105},\ 1 - 1/\sqrt{105})$. Note that the value of f_2 is worse than before despite that DM wants to improve it and hence increase the weight of f_2 up to twice. Someone might think that this is due to the normalization of weight. Therefore, we normalize the weight by $w_1 + w_2 + w_3 = 1$. The original weight normalized in this way is $w_1 = w_2 = w_3 = 1/3$ and the renewed weight by the same normalization is $w_1' = 10/13$, $w_2' = 2/13$, $w_3' = 1/13$. We can observe that w_2' is less than w_2. Now increase the normalized weight w_2 to be greater than 1/3. To this end, set the unnormalized weight $w_1 = 10$, $w_2 = 7$ and $w_3 = 1$. With this new weight, we have a solution $(1 - 10/\sqrt{150},\ 1 - 7/\sqrt{150},\ 1 - 1/\sqrt{150})$. Despite that the normalized weight $w_2'' = 7/18$ is greater than the original one (= 1/3), the obtained solution is still worse than the previous one.

As is readily seen in the above example, it is usually very difficult to adjust the weight in order to obtain a solution as DM wants. Therefore, it seems much better to take the aspiration level of DM rather than the weight as the probe. Interactive multi-objective programming techniques based on aspiration levels have been developed so that the drawbacks of the traditional goal programming may be overcome. In the following section, we shall discuss the satisficing trade-off method developed by the author (Nakayama 1984) as one of them.

3 Satisficing Trade-off Method

In the aspiration level approach, the aspiration level at the k-th iteration $\overline{f}^k$ is modified as follows:

$$\overline{f}^{k+1} = T \circ P(\overline{f}^k) \tag{3.1}$$

Here, the operator P selects the Pareto solution nearest in some sense to the given aspiration level $\overline{f}^k$. The operator T is the trade-off operator which changes the k-th aspiration level $\overline{f}^k$ if DM does not compromise with the shown solution $P(\overline{f}^k)$. Of course, since $P(\overline{f}^k)$ is a Pareto solution, there exists no feasible solution which makes all criteria better than $P(\overline{f}^k)$, and thus DM has to trade-off among criteria if he wants to improve some of criteria. Based on this trade-off, a new aspiration level is decided as $T \circ P(\overline{f}^k)$. Similar process is continued until DM obtains an agreeable solution. This idea is implemented in DIDASS (Wierzbicki 1981 and Grauer *et al.* 1984) and the satisficing trade-off method (Nakayama 1984). In particular, the satisficing trade-off method provides several devices which make the trade-off analysis easier by using the sensitivity analysis and parametric optimization techniques in traditional mathematical programming.

3.1 On The Operation P

The operation which gives a Pareto solution $P(\overline{f}^k)$ nearest to $\overline{f}^k$ is performed by some auxiliary scalar optimization. It has been shown in Sawaragi-Nakayama-Tanino (1985) and Wierzbicki (1986) that the only one scalarization technique, which provides any Pareto solution regardless of the structure of problem, is of the Tchebyshev norm type. However, the scalarization function of Tchebyshev norm type yields not only a Pareto solution but also a weak Pareto solution. Since weak Pareto solutions have a possibility that there may be another solution which improves a critera while others being fixed, they are not necessarily "*efficient*" as a solution in decision making. In order to exclude weak Pareto solutions, the following scalarization function of the augmented Tchebyshev type can be used:

$$\max_{1\le i\le r} w_i(f_i(x)-\overline{f}_i)+\alpha\sum_{i=1}^{r} w_i f_i(x). \tag{3.2}$$

where α is usually set a sufficiently small positive number, say 10^{-6}.

Theorem 3.1 (Nakayama-Tanino 1994) For arbitrary $w \geq 0$ and $\alpha > 0$, $\hat{x} \in X$ minimizing (3.2) is a properly Pareto optimal solution to (MOP). Conversely, if $\hat{x}$ is a properly Pareto optimal solution to (MOP), then there exist $w > 0$, $\alpha > 0$ and $\overline{f}$ such that $\hat{x}$ minimizes (3.2) over X.

The weight w_i is usually given as follows: Let f_i^* be an ideal value which is usually given in such a way that $f_i^* < \text{Min}\{f_i(x)|\ x \in X\}$, and let f_{*i} be a nadir value which is usually given by

$$f_{*i} = \max_{1\le j\le r} f_i(x_j^*) \tag{3.3}$$

where

$$x_j^* = arg\ \min_{x\in X} f_j(x). \tag{3.4}$$

For this circumstance, we set

$$w_i^k = \frac{1}{\overline{f}_i^k - f_i^*} \tag{3.5}$$

or

$$w_i^{k'} = \frac{1}{f_{*i} - f_i^*}. \tag{3.6}$$

The minimization of (3.2) with (3.5) or (3.6) is usually performed by solving the following equivalent optimization problem, because the original one is not smooth:

(Q) Minimize $z + \alpha\sum_{i=1}^{r} w_i f_i(x)$

subject to

$$w_i^k(f_i(x)-\overline{f}_i^k) \leq z \tag{3.7}$$

$$x \in X.$$

Remark 3.1 Note the weight (3.5) depends on the k-th aspiration level, while the one by (3.6) is independent of aspiration levels. The difference between solutions to (Q) for these two kinds of weight can be illustrated in Fig. 3.1. In the auxiliary min-max problem (Q) with the weight by (3.5), $\overline{f}_i^k$ in the constraint (3.7) may be replaced with f_i^* without any change in the solution. For we have

$$\frac{f_i(x)-f_i^*}{\overline{f}_i^k - f_i^*} = \frac{f_i(x)-\overline{f}_i^k}{\overline{f}_i^k - f_i^*} + 1.$$

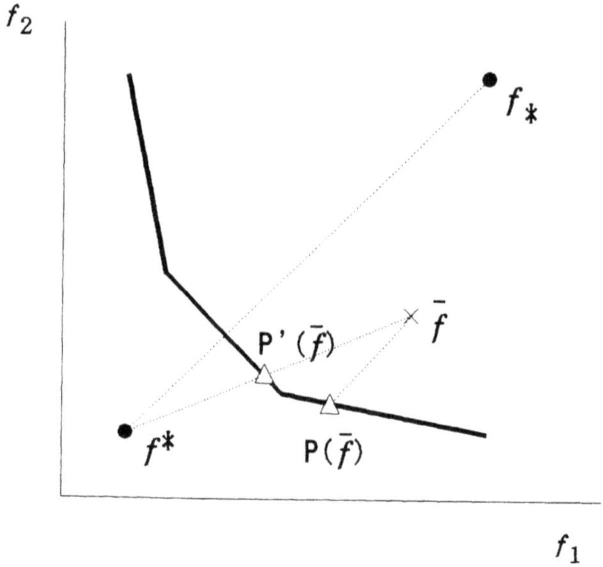

Fig. 3.1 The difference between solutions associated with two kinds of weight

3.2 On The Operation T

In cases DM is not satisfied with the solution for $P(\overline{\boldsymbol{f}}^k)$, he/she is requested to answer his/her new aspiration level $\overline{\boldsymbol{f}}^{k+1}$. Let x^k denote the Pareto solution obtained by projection $P(\overline{\boldsymbol{f}}^k)$, and classify the objective functions into the following three groups:

(i) the class of criteria which are to be improved more,
(ii) the class of criteria which may be relaxed,
(iii) the class of criteria which are acceptable as they are.

Let the index set of each class be denoted by I_I^k, I_R^k, I_A^k, respectively. Clearly, $\overline{f}_i^{k+1} < f_i(x^k)$ for all $i \in I_I^k$. Usually, for $i \in I_A^k$, we set $\overline{f}_i^{k+1} = f_i(x^k)$. For $i \in I_R^k$, DM has to agree to increase the value of $\overline{f}_i^{k+1}$. It should be noted that an appropriate sacrifice of f_j for $j \in I_R^k$ is needed for attaining the improvement of f_i for $i \in I_I^k$.

Example 3.1 Consider the same problem as in Example 2.1: Let $y_1 = f_1(x)$, $y_2 = f_2(x)$ and $y_3 = f_3(x)$, and let the feasible region in the objective space be given by

$$\{(y_1, y_2, y_3)|\ (y_1 - 1))^2 + (y_2 - 1)^2 + (y_3 - 1)^2 \leq 1\}.$$

Suppose that the ideal point is $(y_1^*,\ y_2^*,\ y_3^*) = (0, 0, 0)$, and the nadir point is $(y_{*1},\ y_{*2},\ y_{*3}) = (1, 1, 1)$. Therefore, using (3.6) we have $w_1 = w_2 = w_3 = 1.0$.

Let the first aspiration level be $(\overline{y}_1^1, \overline{y}_2^1, \overline{y}_3^1) = (0.4, 0.4, 0.4)$. Then the solution to (Q) is $(y_1^1, y_2^1, y_3^1) = (0.423, 0.423, 0.423)$. Now suppose that DM wants to decrease the value of f_1 a lot more and that of f_2 a little more, and hence modify the aspiration level into $\overline{y}_1^2 = 0.35$ and $\overline{y}_2^2 = 0.4$. Since the present solution $(y_1^1, y_2^1, y_3^1) = (0.423, 0.423, 0.423)$ is already Pareto optimal, it is impossible to improve all of criteria. Therefore, suppose that DM agrees to relax f_3, and with its new aspiration level of $\overline{y}_3^2 = 0.5$. With this new aspiration level, the solution to (Q) is $(y_1^2, y_2^2, y_3^2) = (0.359, 0.409, 0.509)$. Although the obtained solution does not attain the aspiration level of f_1 and f_2 a little bit, it should be noted that the solution is improved more than the previous one. The reason why the improvement of f_1 and f_2 does not attain the wish of DM is that the amount of relaxation of f_3 is not much enough to compensate for the improvement of f_1 and f_2. In the satisficing trade-off method, DM can find a satisfactory solution easily by making the trade-off analysis deliberately. To this end, it is also possible to use the sensitivity analysis in mathematical programming. (Refer to the following automatic trade-off or exact trade-off). We have observed in the previous section that it is difficult to adjust weights for criteria so that we may get a desirable solution in the goal programming. However, the aspiration level can lead DM to his/her desirable solution easily in many practical problems.

3.3 Automatic Trade-off

It is of course possible for DM to answer new aspiration levels of all objective functions. In practical problems, however, we often encounter cases with very many objective functions, for which DM tends to get tired with answering new aspiration levels for all objective functions. Therefore, it is more practical in problems with very many objective functions for DM to answer only his/her improvement rather than both improvement and relaxation. At this stage, we can use the assignment of sacrifice for f_j $(j \in I_R)$ which is automatically set in the equal proportion to $(\lambda_i + \alpha)w_i$, namely, by

$$\Delta f_j = \frac{-1}{N(\lambda_j + \alpha)w_j} \sum_{i \in I_I} (\lambda_i + \alpha) w_i \Delta f_i, \qquad j \in I_R \tag{3.8}$$

where N is the number of elements of the set I_R, and λ is the Lagrange multiplier associated with the constraints in Problem (Q). The reason why (3.8) is available is given by the following:

Theorem 3.2 (Nakayama 1991-b, Nakayama-Tanino 1994) Let $\tilde{x}$ be a solution to (Q) with $X = \{x |\ g_j(x) \leq 0,\ j = 1, \ldots, m\}$ and $\tilde{f} = f(\tilde{x})$. Suppose that the second order sufficient condition holds; namely,

$$u^T \nabla^2 L(\tilde{x}; \lambda, \mu) u > 0$$

for any $u \neq 0$ such that

$$\nabla f_i(\tilde{x}) u = 0, \quad i = 1, \ldots, r-1$$

and

$$\nabla g_j(\tilde{x})u = 0, \quad j \in J = \{j | \; g_j(\tilde{x}) = 0\}.$$

Suppose also that the vectors $(\nabla f_1(\tilde{x}), -1), \ldots, (\nabla f_r(\tilde{x}), -1)$, $(\nabla g_{k_1}(\tilde{x}), 0)$, $\ldots, (\nabla g_{k_s}(\tilde{x}), 0)$ are linearly independent, where $(k_1, \ldots, k_s)$ is the index set of active constraints of (Q). In addition, suppose that the following strict complementary slackness condition holds:

$$\lambda_i > 0 \quad \text{for any} \quad i \in 1, \ldots, r$$

$$\mu_j > 0 \quad \text{for any} \quad j \in J.$$

Then, with Δf_i $(i = 1, \ldots, r)$ such that $(\tilde{f}_1 + \Delta f_1, \ldots, \tilde{f}_r + \Delta f_r)$ is on the Pareto surface, we have

$$0 = \sum_{i=1}^{r} \lambda_i \Delta f_i + o(\| \Delta f \|)$$

Therefore, under some appropriate condition, $((\lambda_1 + \alpha)w_1, \ldots, (\lambda_r + \alpha)w_r)$ is the normal vector of the tangent hyperplane of the Pareto surface. In particular, in multi-objective linear programming problems, the simplex multipliers corresponding to a nondegenerated solution is the feasible trade-off vector of Pareto surface (Nakayama 1992).

By using the above automatic trade-off method, the burden of DM can be decreased so much in cases that there are a large number of criteria. Of course, if DM does not agree with this quota Δf_j laid down automatically, he/she can modify it in a manual way.

Example 3.2 Consider the same problem as in Example 3.1. Suppose that DM has the solution $(y_1^1, y_2^1, y_3^1) = (0.423, 0.423, 0.423)$ associated with his first aspiration level $(\overline{y}_1^1, \overline{y}_2^1, \overline{y}_3^1) = (0.4, 0.4, 0.4)$. The Lagrange multipliers at this solution is $\lambda_1 = \lambda_2 = \lambda_3 = 0.333$. Now suppose that DM modifies the aspiration level into $\overline{y}_1^2 = 0.35$ and $\overline{y}_2^2 = 0.4$. For the amount of improvement of $|\Delta f_1| = 0.073$ and $|\Delta f_2| = 0.023$, the amount of relaxation of f_3 on the basis of automatic trade-off is $|\Delta f_3| = 0.095$. In other words, the new aspiration level of f_3 should be 0.52. If DM agrees with this trade-off, he/she will have the new Pareto solution $(y_1^2, y_2^2, y_3^2) = (0.354, 0.404, 0.524)$ to the problem (Q) corresponding to the new aspiration level $(\overline{y}_1^2, \overline{y}_2^2, \overline{y}_3^2) = (0.35, 0.4, 0.52)$. It should be noted that the obtained solution is much closer to DM's wish rather than the one in Example 3.1.

Example 3.3 Consider the following multiple objective linear programming problem:

$$\begin{aligned} f_1 &= -2x_1 - x_2 + 25 \rightarrow Min \\ f_2 &= x_1 - 2x_2 + 18 \rightarrow Min \end{aligned}$$

subject to

$$
\begin{array}{rcl}
-x_1 + 3x_2 & \leq & 21 \\
x_1 + 3x_2 & \leq & 27 \\
4x_1 + 3x_2 & \leq & 45 \\
3x_1 + x_2 & \leq & 30 \\
x_1,\ x_2 & \geq & 0
\end{array}
$$

Suppose that the ideal point $\boldsymbol{f}^* = (4,\ 4)$ and the nadir point $\boldsymbol{f}_* = (18,\ 21)$ by using the pay-off matrix based on minimization of each objective function separately. Lettiing the first aspiration level be $\overline{\boldsymbol{f}}^1 = (15,\ 9)$, we have the first Pareto solution $(11.632,\ 4.910)$ by solving the auxiliary min-max problem (Q). This Pareto point in the objective function space is the intercept of the line parallel to the line passing through $\boldsymbol{f}^*$ and $\boldsymbol{f}_*$ with the Pareto surface (curve, in this case). Now we shall consider the following three cases:

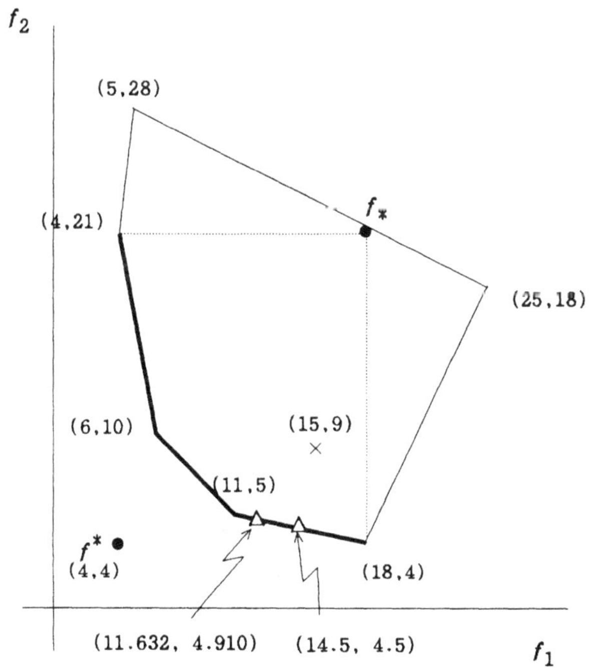

Fig. 3.2 Automatic trade-off (case 1)

<Case 1> Suppose that DM is not satisfied with the obtained Pareto solution, and he/she wants to improve the value of f_2. Let the new aspiration level of f_2 be

4.5. Instead, suppose that DM agrees with some sacrifice of f_1. The new aspiration level of f_1 based on the automatic trade-off is 14.5. Since the automatic trade-off is based on the linear approximation of Pareto surface at the present point, the new aspiration level obtained by the automatic trade-off is itself Pareto optimal in this case as shown in Fig. 3.2.

<**Case 2**> Suppose that DM wants to improve f_1 rather than f_2 at the moment when the first Pareto solution (11.632, 4.910) is obtained. Let the new aspiration level of f_1 that DM desires be 9.0. Then the new aspiration level by the automatic trade-off is $\overline{\boldsymbol{f}}^2 = (9.0,\ 5.286)$, and is not Pareto optimal. Solving the auxiliary min-max problem (Q) with the new aspiration level, in this case, we have the new Pareto solution (9.774, 6.226) as in Fig. 3.3. Since the improvement which DM desires is not so large after solving the min-max problem (Q) with an aspiration level in many practical cases, the new aspiration level produced by automatic trade-off based on the linear approximation of Pareto surface is close to the Pareto surface. Therefore, the satisficing trade-off method using the automatic trade-off yields the desirable solution usually only in a few iterations.

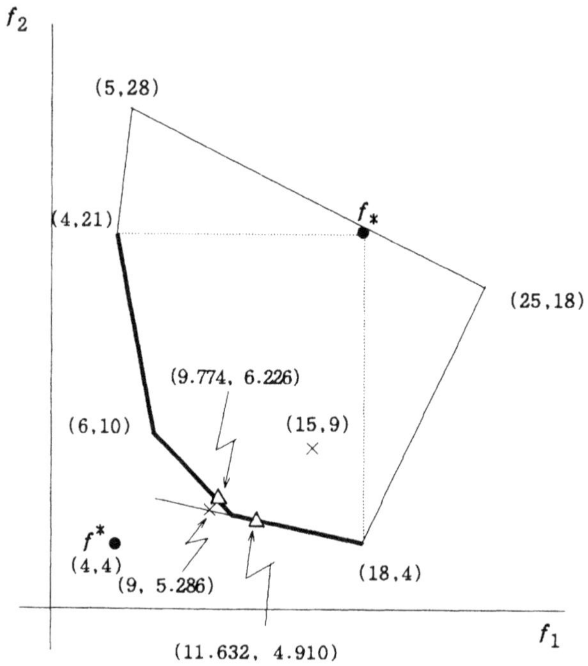

Fig. 3.3 Automatic trade-off (case 2)

<**Case 3**> Suppose that DM wants to make f_1 less than 9.0 absolutely at the moment when the first Pareto solution (11.632, 4.910) is obtained. In this case, we have to treat f_1 as the constraint

$$f_1(x) \leq 9.0.$$

As will be shown in the subsection 3.5 below, the interchange between objectives and constraints can be made so easily in the formulation of auxiliary min-max problem. (We can treat the criteria as DM wishes between objectives and constraints by adjusting one parameter β in the min-max problem.)

3.4 Exact Trade-off

In linear or quadratic cases, we can evaluate the exact trade-off in an extended form of the automatic trade-off stated above. This implies that we can calculate exactly the amount of relaxation such that the new aspiration level is on the Pareto surface (Nakayama, 1992). The main idea in it is that the parametric optimization technique is used in stead of the simple sensitivity analysis. Using this technique, a new Pareto solution can be obtained without solving the auxiliary sclarized optimization problem again. This implies that we can obtain the new solution very quickly. Therefore, using some graphic presentation as computer outputs, DM can see the trade-off among criteria in a dynamic way, e.g. as an animation. This makes DM's judgement easier.

3.5 Interchange between Objectives and Constraints

In the formulation of the auxiliary scalarized optimization prolem (Q), change the right hand side of the equation (3.7) into $\beta_i z$, namely

$$w_i(f_i(x) - \overline{f}_i) \leq \beta_i z. \tag{3.9}$$

As is readily seen, if $\beta_i = 1$, then the function f_i is considered to be an objective function, for which the aspiration level $\overline{f}_i$ is not necessarily attained, but the level of f_i should be better as much as possible. On the other hand, if $\beta_i = 0$, then f_i is considered to be a constraint function, for which the aspiration level $\overline{f}_i$ should be guaranteed. In many practical problems, there is almost no cases in which we consider the role of objective and constraint fixed from the beginning, but usually we want to interchange them depending on the situation. Using the formula (3.9), this can be done very easily (Korhonen 1987). In addition, if the value of β_i is set in the middle of 0 and 1, f_i can play a role in the middle of objective and constraint which is neither a complete objective nor a complete constraint (Kamenoi *et al.* 1992). This is also very effective in many practical problems.

3.6 Remarks on Trade-off for Objective Functions with 0-Sensitivity

Since α is sufficiently small like 10^{-6} and $\lambda_1 + \ldots \lambda_r = 1$, we can consider in many cases

$$(\lambda_i + \alpha)w_i \simeq \lambda_i w_i.$$

When $\lambda_j = 0$ and f_j is not to be improved, we set $\Delta f_j = 0$ in the automatic trade-off. Therefore, unless we select at least one f_j with $\lambda_j \neq 0$ as an objective function to be relaxed, we can not attain the improvement that DM wishes.

Since $(\lambda_i + \alpha)w_i$ (or approximately, $\lambda_i w_i$) can be regarded to provide the sensitivity information in the trade-off, $\lambda_i = 0$ means that the objective function f_i does not contribute to the trade-off among the objective functions. In other words, since the trade-off is the negative correlation among objective functions, $\lambda_i = 0$ means that f_i has the positive correlation with some other objective functions. Therefore, if all objective functions to be relaxed, $f_j \ (j \in I_R)$, have $\lambda_j = 0 \ (j \in I_R)$, then they can not compensate for the improvement which DM wishes, because they are affected positively by some of objective functions to be improved.

Example 3.4 Consider the following problem:

$$\text{Minimize} \quad (f_1, \ f_2, \ f_3) = (x_1, \ x_2, \ x_3)$$

subject to

$$\begin{aligned} x_1 + x_2 + x_3 &\geq 1 \\ x_1 \qquad - x_3 &\geq 0 \\ x_1, \ x_2, \ x_3 &\geq 0 \end{aligned}$$

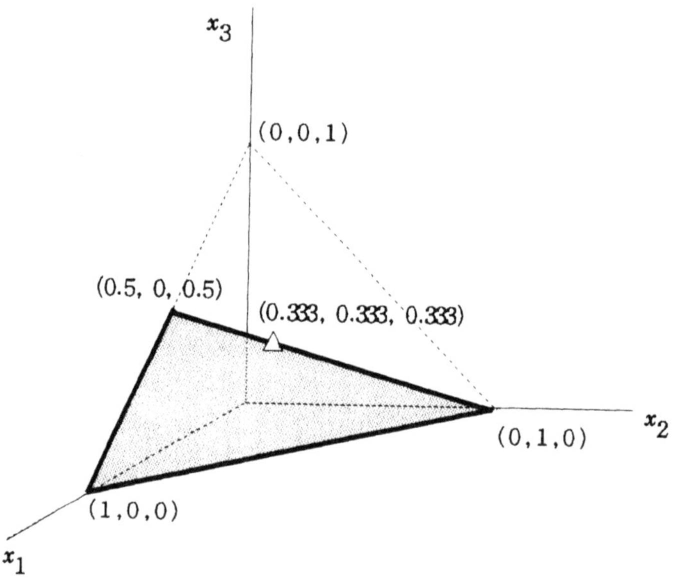

Fig. 3.4 A case of automatic trade-off with 0-sensitivity

The ideal point is (0, 0, 0) and the nadir point is (1.0, 1.0, 0.5). For the first aspiration level (0.2, 0.2, 0.4), we have a Pareto value (0.333, 0.333, 0.333) and the corresponding simplex multiplier $(\lambda_1, \lambda_2, \lambda_3) = (2/3, 1/3, 0)$ as shown in Fig. 3.4. Suppose that DM wants to improve f_1 and f_2, and sets their new aspiration levels 0.2 and 0.3, respectively. Since the relaxation $\Delta f_3 = 0$ by the automatic trade-off, the new aspiration level becomes (0.2, 0.3, 0.333). Associated with the new aspiration level, we have the Pareto value (0.3, 0.4, 0.3), in which neither f_2 is improved nor f_3 is relaxed. This is because that the objective functions f_1 and f_3 has a positive correlation along the edge of Pareto surface at the point (0.333, 0.333, 0.333), while f_1 and f_2 have trade-off relation with each other there. As a result, though f_3 was considered to be relaxed, it was affected strongly by f_1 and hence improved. On the other hand, despite that f_3 was considered to be improved, it was relaxed finally. This is due to the fact that the objective function to be relaxed is only f_3 despite that λ_3 is 0, and the fact that we did not consider that f_3 has positive correlation with f_1. This example suggests that we should make the trade-off analysis deliberately seeing the value of simplex multiplier (Lagrange multiplier, in nonlinear cases). Like this, the satisficing trade-off method makes the DM's trade-off analysis easier by utilizing the information of sensitivity.

3.7 Relationship to Fuzzy Mathematical Programming

In the aspiration level approach to multi-objective programming such as the satisficing trade-off method, the wish of DM is attained by adjusting his/her aspiration level. In other words, this means that the aspiration level approach can deal with the fuzziness of right hand side value in traditional mathematical programming as well as the total balance among the criteria. There is another method, namely fuzzy mathematical programming, which treat the fuzziness of right hand side value of constraint in traditional mathematical programming. In the following, we shall discuss the relationship between the satisficing trade-off method and the fuzzy mathematical programming.

For simplicity, consider the following problem:

$$
\begin{array}{lll}
(\mathrm{F}) & \text{Maximixe} & f_0(x) \\
 & \text{subject to} & f_1(x) = \overline{f}_1
\end{array}
$$

Suppose that the right hand side value $\overline{f}_1$ is not needed to meet so strictly, but that it is fuzzy. The membership function for the criterion f_1 is usually given as in Fig. 3.5. Since our aim is to maximize this membership function, we can adopt the following function without change in the solution:

$$m_1(x) = \min \{(\overline{f}_1 - f_1(x))/\epsilon + 1, \ -(\overline{f}_1 - f_1(x))/\epsilon + 1\},$$

where ϵ is a parameter representing the admissible error for the target $\overline{f}_1$.

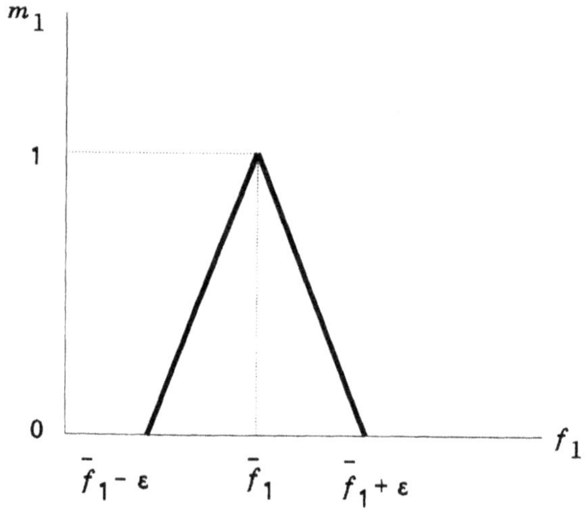

Fig. 3.5 Membership function for f_1 in (F)

Now, the problem (F) is reduced to a kind of multi-objective optimization in which f_0 and m_1 should be maximized. Then a membership function for maximization of f_0 is usually defined with its aspiration level $\overline{f}_0$. For example,

$$m_0'(x) = \min\{-(\overline{f}_0 - f_0(x))/(f_0^* - f_{0*}) + 1, \;\; 1\},$$

However, if we maximize the above m_0' as it is, the solution will be merely the one to the satisficing problem for which f_0 is to be just greater than $\overline{f}_0$. As was stated in the previous section, we shall use the following function in stead of m_0' in order to assure the Pareto optimality of the solution:

$$m_0(x) = -(\overline{f}_0 - f_0(x))/(f_0^* - f_{0*}) + 1.$$

Finally, our problem is to maximize both m_0 and m_1, which is usually reduced to the following problem:

$$\begin{array}{ll} \text{Minimize} & z \\ \text{subject to} & \end{array}$$

$$\begin{aligned} (\overline{f}_0 - f_0(x))/(f_0^* - f_{0*}) - 1 &\leq z \\ (\overline{f}_1 - f_1(x))/\epsilon - 1 &\leq z \\ -(\overline{f}_1 - f_1(x))/\epsilon - 1 &\leq z \end{aligned}$$

Now, one may see some similarity between the above formulation and the one in the satisficing trade-off method. In the satisficing trade-off method, the objective function with target such as $f_1 \rightarrow \overline{f}_1$ is usually treated as two objective functions, $f_1 \rightarrow Max$ and $f_1 \rightarrow Min$. Under this circumstance, suppose that for $f_1 \rightarrow Max$ we set the ideal value $f_1^* = \overline{f}_1$, the nadir value $f_{1*} = \overline{f}_1 - \epsilon$ and the aspiration level $\overline{f}_1 - \epsilon$; for $f_1 \rightarrow Min$ we set the ideal value $f_1^* = \overline{f}_1$, the nadir value $f_{1*} = \overline{f}_1 + \epsilon$ and the aspiration level $\overline{f}_1 + \epsilon$. Then the treatment of f_1 is the same between the above formulation and the satisficing trade-off method.

However, usually in the satisfcing trade-off method, we set f_1^* and f_{1*} based on the pay-off matrix (i.e., different from $\overline{f}_1$ and $\overline{f}_1 \pm \epsilon$, respectively). Hence, we do not contain ϵ in the denominator of constraints in the min-max problem, because we make the trade-off analysis by adjusting ϵ rather than the target $\overline{f}_1$; for example, using (3.6) the constraint for f_1 in the min-max problem is given by

$$\begin{aligned} (\overline{f}_1 - \epsilon - f_1(x))/(f_1^* - f_{1*}) &\leq z \\ -(\overline{f}_1 + \epsilon - f_1(x))/(f_1^* - f_{1*}) &\leq z \end{aligned}$$

With this formulation, even if DM wants $\epsilon = 0$ and if there is no solution to $f_1(x) = \overline{f}_1$, we can get a solution approximate to $f_1(x) = \overline{f}_1$ as much as possible. In the fuzzy mathematical programming, however, if $\epsilon = 0$, then we have a crisp constraint $f_1(x) = \overline{f}_1$, and we sometimes have no feasible solution to it.

Finally as a result, we can see that the satisficing trade-off method deals with the fuzziness of right hand side value of constraint automatically and can effectively treat problems for which fuzzy mathematical programming provides no solution. Due to this reason, we can conclude that it is better to formulate the given problem as a multi-objective optimization from the beginning and to solve it by the aspiration level approach such as the satisficing trade-off method

4 Applications

Interactive multi-objective programming methods have been applied to a wide range of practical problems. Good examples in engineering applications can be seen in Eschenauer *et al.* (1990). The author himself also has applied to several real problems:

1. blending
   (a) feed formulation for live stock (Nakayama *et al.* 1992)
   (b) plastic materials (Nakayama *et al.* 1986)
   (c) cement production (Nakayama 1991-a)

   (d) portfolio (Nakayama 1989)

2. design

   (a) camera lens

   (b) erection management of cable-stayed bridge (Furukawa *et al.* 1986)

3. route search

   (a) car navigation system (Nakayama *et al.* submitted)

4. planning

   (a) scheduling of string selection in steel manufacturing (Ueno *et al.* 1990)

   (b) long term planning of atomic power plants

In the following, some of examples are introduced briefly.

4.1 Feed Formulation for Live Stock

Stock farms in Japan are modernized recently. Above all, the feeding system in some farms is fully controled by computer: Each cow has its own place to eat which has a locked gate. And each cow has a key on her neck, which can open the corresponding gate only. Every day, on the basis of ingredient analysis of milk and/or of the growth situation of cow, the appropriate blending ratio of materials from several viewpoints should be made.

There are about 20-30 kinds of raw materials for feed in cow farms such as corn, cereals, fish meal, etc.

About 10 criteria are usually taken into account for feed formulation of cow:

- cost
- nutrition
  - protein
  - TDN
  - cellulose
  - calcium
  - magnesium

- etc.

- stock amount of materials
- etc.

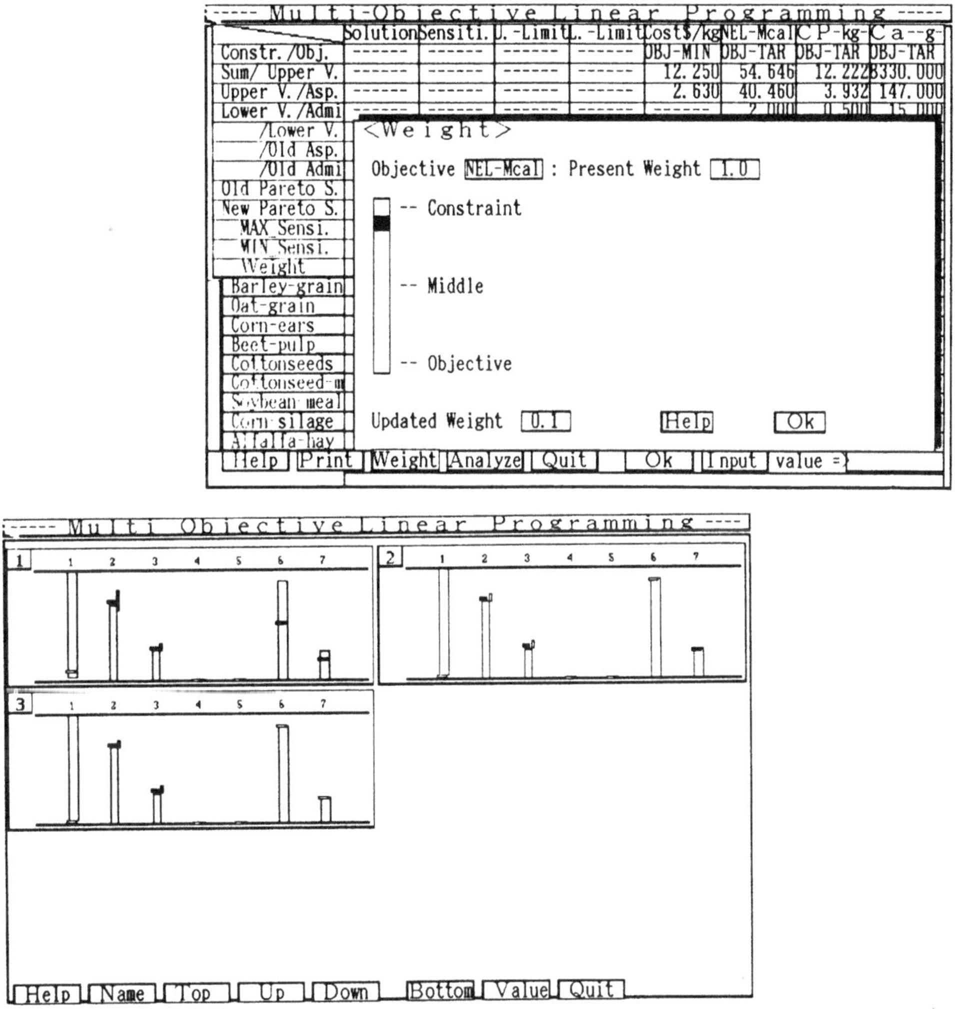

Fig. 4.1 A Phase of F-STOM

This feeding problem is well known as the diet problem from the beginning of the history of mathematical programming, which can be formulated as the traditional linear programming. In the traditional mathematical programming, the solution often occurs on the boundary of constraints. In many cases, however, the right hand side value of constraint such as neutrition needs not to be satisfied rigidly. Rather, it seems to be natural to consider that a criterion such as neutrition is an objective function whose target has some allowable range. As was seen in the previous section, the satisficing trade-off method deals well with the fuzziness of target of such an objective function. The author and others have developed a software for feed formulation using the satisficing trade-off method, called F-STOM (Feed formulation by Satisficing Trade-Off Method) (Nakayama *et al.* 1993). This software is being distributed to live stock farmers and feed companies in Japan through an association of live stock systems.

4.2 Erection Management of Cable Stayed Bridge

In erection of cable stayed bridge, the following criteria are considered for accuracy control (Furukawa *et al.* 1986):

i. residual error in each cable tension,

ii. residual error in camber at each node,

iii. amount of shim adjustment for each cable,

iv. number of cables to be adjusted.

Since the change of cable rigidity is small enough to be neglected with respect to shim adjustment, both the residual error in each cable tension and that in each camber are linear functions of amount of shim adjustment. Let us define n as the number of cable in use, $\Delta T_i \quad (i = 1, \ldots, n)$ as the difference between the designed tension values and the measured ones, and x_{ik} as the tension change of i-th cable caused from the change of the k-th cable length by a unit. The residual error in cable tension caused by the shim adjustment is given by

$$p_i = |\Delta T_i - \sum_{k=1}^{n} x_{ik} \Delta l_k| \qquad (i = 1, \ldots, n)$$

Let m be the number of nodes, $\Delta z_j \quad (j = 1, \ldots, m)$ the difference between the designed camber values and the measured ones, and y_{jk} the camber change at j-th node caused from the change of the k-th cable length by a unit. Then the residual error in the camber caused by the shim adjustments of $\Delta l_1, \ldots, \Delta l_n$ is given by

$$q_j = |\Delta Z_j - \sum_{k=1}^{n} y_{jk} \Delta l_k| \qquad (j = 1, \ldots, m)$$

In addition, the amount of shim adjustment can be treated as objective functions of

$$r_i = |\Delta l_i| \qquad (i = 1, \ldots, n)$$

And the upper and lower bounds of shim adjustment inherent in the structure of the cable anchorage are as follows;

$$\Delta l_{Li} \leq \Delta l_i \leq \Delta l_{Ui} \qquad (i = 1, \ldots, n).$$

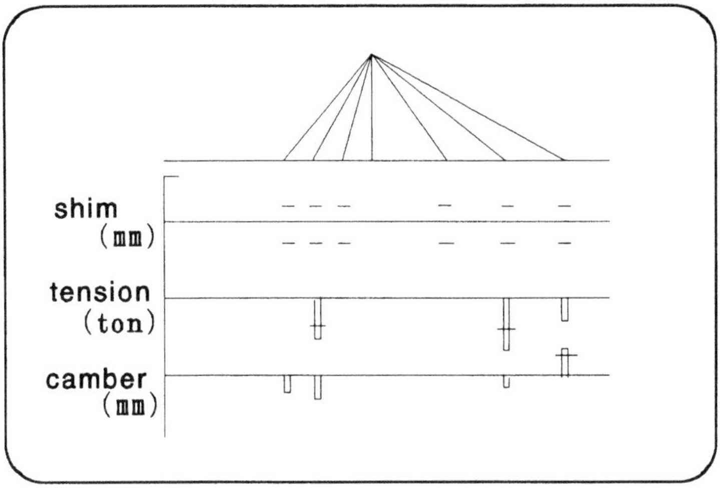

Fig. 4.2 A phase of the proposed erection management system

Fig.4.2 shows one phase of erection management system of cable stayed bridge using the satisficing trade-off method. The residual error of each criterion and the amount of shim adjustment are represented by graphs. The aspiration level is inputted by a mouse on the graph. After solving the auxiliary min-max problem, the Pareto solution according to the aspiration level is represented by a graph in a similar fashion. This procedure is continued until the designer can obtain a desirable shim-adjustment. This operation is very easy for the designer, and the visual information on trade-off among criteria is user-friendly. The software was used for real bridge construction, say, Tokiwa Great Bridge (Ube City) and Karasuo Harp Bridge (Kita-Kyusyu City) in 1992.

4.3 An Interactive Support System for Bond Trading

In portfolio selection problems, many companies are now widely trying to use mathematical analysis for bond trading. In this subsection, some bond trading problem is formulated as a kind of multi-objective fractional problem. It will be seen in the following that the satisficing trade-off method can be effectively applied to such a porfolio problem.

Bond traders are facing almost every day a problem which bonds and what amount they should sell and/or buy in order to attain their customers' desires. The economic environment is changing day by day, and sometimes gives us a drastic change. Bond traders have to take into account many factors, and make their decisions very rapidly and flexibly according to these changes. The number of bonds to be considered is usually more than 500, and that of criteria is about ten as will be shown later. The amount of trading is usually more than 100 million yen, and hence even a slight difference of portfolio combination might cause a big difference to profit or loss. This situation requires some effective method which helps bond traders following faithfully their value-judgment on a real time basis not only mathematically but also in such a way that their intuition fostered by their experiences can be taken in.

Bond portfolio problems are a kind of blending problems. Therefore, mathemetical programming approach can be used very effectively. However, the traditional mathematical programming approach with a single objective function can not take in the value-judgment and intuition of bond traders so easily in a flexible manner for the changes of environment. We shall show that the satisficing trade-off method fits to this purpose.

4.3.1 Mathematical Formulation for Bond Portfolio Problems

We shall follow the mathematical model given by Konno and Innori (1987). Assume that an investor holds u_j units of bonds B_j, $j = 1, \ldots, N$. Associated with B_j, we have the following indices:

c_j: coupon to be paid at a fixed rate (yen/bond/year)
f_j: principal value to be refunded at maturity (yen/bond)
p_j: present price in the market (yen/bond)
t_j: maturity (number of years until its principal value is refunded)

Returns from bonds are the income from coupon and the capital gain due to price increase. Bond portfolio problems are to determine which bonds and what amount the investor should sell and/or buy taking into account many factors, say, expected returns and risk, the time needs money for another investment, and so on. Therefore, we have the following criteria for each bond B_j:

[I] returns

(i) direct yield (short term index of return)

$$\gamma_j = \frac{c_j}{p_j}$$

(ii) effective yield

$$\nu_j = \frac{\left[c_j\{(1+\alpha)^{t_j} - 1\}/\alpha + f_j\right]^{1/t_j}}{p_j} - 1$$

where α is the interest rate.

[II] risk

(iii) price variation

$$\pi_j = \frac{t_j}{1 + \mu_j t_j}, \quad j = 1, \ldots, N$$

Let x_j $(j = 1, ..., n_1)$ and X_k $(k = 1, ..., n_2)$ denote the number of bonds to be sold and to be purchased, respectively. Then S_0 and S_1 represent respectively the total quantity of bonds and the total value of bonds after the transaction. Namely,

$$\begin{aligned} S_0 &= \sum_{j=1}^{N} u_j - \sum_{j=1}^{n_1} x_j + \sum_{k=1}^{n_2} X_k \\ S_1 &= \sum_{j=1}^{N} p_j u_j - \sum_{j=1}^{n_1} p_j x_j + \sum_{k=1}^{n_2} P_k X_k . \end{aligned}$$

In addition, we set

$$S_2 = \sum_{j=1}^{N} p_j t_j u_j - \sum_{j=1}^{n_1} p_j t_j x_j + \sum_{k=1}^{n_2} P_k T_k X_k$$

Then the average for each index is taken as an objective function:

(i′) average direct yield

$$F_1 = \left(\sum_{j=1}^{N} \gamma_j p_j u_j - \sum_{j=1}^{n_1} \gamma_j p_j x_j + \sum_{k=1}^{n_2} \gamma_k P_k X_k \right) / S_1$$

(ii′) average effective yield

$$F_2 = \left(\sum_{j=1}^{N} \nu_j p_j t_j u_j - \sum_{j=1}^{n_1} \nu_j p_j t_j x_j + \sum_{k=1}^{n_2} \nu_k P_k T_k X_k \right) / S_2$$

(iii′) average price variation

$$F_3 = \left(\sum_{j=1}^{N} \pi_j u_j - \sum_{j=1}^{n_1} \pi_j x_j + \sum_{k=1}^{n_2} \pi_k X_k \right) / S_0$$

Our constraints are divided into soft constraints and hard constraints:

(iv) average unit price

$$F_4 = \left(\sum_{j=1}^{N} p_j u_j - \sum_{j=1}^{n_1} p_j x_j + \sum_{k=1}^{n_2} P_k X_k \right) / S_0$$

(v) average maturity

$$F_5 = \left(\sum_{j=1}^{N} t_j u_j - \sum_{j=1}^{n_1} t_j x_j + \sum_{k=1}^{n_2} T_k X_k \right) / S_0$$

(vi) specification of time of coupon to be paid

$$F_6 = \sum_{i \in I_1} x_i / S_0$$
$$F_7 = \sum_{i \in I_2} x_i / S_0$$

where I_m is the set of indices of bonds whose coupon are paid at the time t_m.

[III] hard constraints

(vii) budget constraints:

$$\sum_{j=1}^{n_1} -p_j x_j + \sum_{k=1}^{n_2} P_k X_k \leq C$$
$$\sum_{k=1}^{n_2} P_k X_k \geq C$$

(viii) specification of bond

$$l_j \leq x_j \leq u_j, \quad j = 1, ..., n_1$$
$$L_k \leq X_k \leq U_k, \quad k = 1, ..., n_2$$

For this kind of problem, the satisficing trade-off method can be effectively applied. Then we have to solve a linear fractional min-max problem. In the following subsection, we shall give a brief review of the method for solving it.

4.3.2 An Algorithm for Linear Fractional Min-Max Problems

Let each objective function in our bond trading problem be of the form $F_i(x) = p_i(x)/q_i(x)$ $(i = 1, \ldots, r)$ where p_i and q_i are linear in x. Then since

$$F_i^* - F_i(x) = \frac{F_i^* q_i(x) - p_i(x)}{q_i(x)} := \frac{f_i(x)}{g_i(x)}$$

the auxiliary min-max problem (Q) becomes a kind of linear fractional Min-Max problem. For this kind of problem, several methods have been developed: Here we shall use a Dinkelbach type algorithm (Borde-Crouzeix 1987, Ferland-Potvin 1985) as is stated in the following:

Step 1: Let $x^0 \in X$. Set $\theta^0 = \max_{1 \leq i \leq r} f_i(x^0)/g_i(x^0)$ and $k = 0$.

Step 2: Solve the problem

$$(P_k) \qquad T_k(\theta^k) = \min_{x\in X} \max_{1\le i\le r} (f_i(x) - \theta^k g_i(x))/g_i(x^k)$$

Let x^{k+1} be a solution to (P_k).

Step 3: If $T_k(\theta^k) = 0$ then stop: θ^k is the optimal value of the given min-max Problem, and x^{k+1} is the optimal solution.

Step 4: If not, take $\theta^{k+1} = \max_{1\le i\le r} f_i(x^{k+1})/g_i(x^{k+1})$. Replace k by $k+1$ and go to Step 2.

Note that thc problem (P_k) is the usual linear min-max problem. Therrefore, we can obtain its solution by solving the following equivalent problem in a usual manner:

$$(Q_k) \qquad \text{Minimize} \quad z$$
$$\text{subject to}$$

$$\frac{f_i(x) - \theta^k g_i(x)}{g_i(x^k)} \le z, \quad i = 1, ..., r$$

4.3.3 An Experimental Result

A result of our experiments is shown below: The problem is to decide a bond portfolio among 37 bonds selected from the market. The holding bonds are $x(1) = 5000$, $x(9) = 1000$, $x(13) = 2500$, $x(17) = 4500$, $x(19) = 5500$, $x(21) = 6000$, $x(23) = 5200$, $x(25) = 4200$, $x(27) = 3200$ and $x(37) = 3800$. The experiment was performed by a worker of a security company in Japan who has a carreer of acting as a bond trader. For the first aspiration level $F1 = 5.8$, $F2 = 6.8$, $F3 = 0.13$ with constraints $F4 \le 103.0$, $4.00 \le F5 \le 5.00$, $0.20 \le F6$ and $0.20 \le F7$, we have a result as shown in Table 4.1.

The asterisk of F4 to F7 implies soft constraints. In our system, we can change objective function into soft constraints and vice versa. Here, the bond trader changed F2 (effective yield) into a soft constraint, and F4 (unit price) into an objective function. Then under the modified aspiration level by trade-off, the obtained result is given by Table 4.2.

The result was satisfactory to the testee. He recognized that this method (software) is very easy to use and flexible for variation of the desire of investers. Transaction cost should be taken into account in the future.

Table 4.1 Result for the first aspiration level

		Pareto sol.	Asp.Level (Target Range)	Lowest	Highest	Sensitivity
	F1 (max)	5.8335	5.8000	5.4788	5.9655	0.0133
	F2 (max)	6.8482	6.8000	6.7165	6.8698	0.0000
	F3 (min)	0.1292	0.1300	0.1262	0.1325	1.0000
*	F4	102.7157	$F4 \leq 103.00$			
*	F5	4.0000	$4.00 \leq F5 \leq 5.00$			
*	F6	0.2000	$0.20 \leq F6$			
*	F7	0.2000	$0.20 \leq F7$			

x(1)=	1042.0643	x(2)=	400.0000
x(3)=	400.0000	x(4)=	200.0000
x(5)=	0.0000	x(6)=	400.0000
x(7)=	0.0000	x(8)=	0.0000
x(9)=	547.3548	x(10)=	0.0000
x(11)=	0.0000	x(12)=	0.0000
x(13)=	2500.0000	x(14)=	200.0000
x(15)=	0.0000	x(16)=	200.0000
x(17)=	4500.0000	x(18)=	0.0000
x(19)=	5321.9573	x(20)=	0.0000
x(21)=	6000.0000	x(22)=	0.0000
x(23)=	5200.0000	x(24)=	0.0000
x(25)=	4200.0000	x(26)=	0.0000
x(27)=	3200.0000	x(28)=	274.6025
x(29)=	400.0000	x(30)=	200.0000
x(31)=	0.0000	x(32)=	200.0000
x(33)=	400.0000	x(34)=	400.0000
x(35)=	0.0000	x(36)=	0.0000
x(37)=	3800.0000		

Table 4.2 Result for the modified aspiration level

		Pareto sol.	Asp.Level (Target Range)	Lowest	Highest	Sensitivity
	F1 (max)	5.8629	5.9000	5.8193	5.9608	0.0000
*	F2 (max)	6.8482	6.8482 ≤ F2			
	F3 (min)	0.1302	0.1292	0.1291	0.1322	1.0000
	F4 (min)	102.3555	102.1000	102.0676	102.7228	0.0043
*	F5	4.0000	4.00 ≤ F5 ≤ 5.00			
*	F6	0.2000	0.20 ≤ F6			
*	F7	0.2000	0.20 ≤ F7			

x(1)=	0.0000	x(2)=	400.0000
x(3)=	400.0000	x(4)=	200.0000
x(5)=	0.0000	x(6)=	400.0000
x(7)=	0.0000	x(8)=	0.0000
x(9)=	139.4913	x(10)=	0.0000
x(11)=	0.0000	x(12)=	0.0000
x(13)=	2500.0000	x(14)=	381.5260
x(15)=	0.0000	x(16)=	200.0000
x(17)=	4500.0000	x(18)=	0.0000
x(19)=	5277.1577	x(20)=	0.0000
x(21)=	6000.0000	x(22)=	0.0000
x(23)=	5200.0000	x(24)=	0.0000
x(25)=	4200.0000	x(26)=	0.0000
x(27)=	3200.0000	x(28)=	14.8920
x(29)=	400.0000	x(30)=	200.0000
x(31)=	0.0000	x(32)=	200.0000
x(33)=	400.0000	x(34)=	400.0000
x(35)=	400.0000	x(36)=	222.7743
x(37)=	3800.0000		

5 Concluding Remarks

Recently, much attention has been being paid to the intelligent information process by computers. Such intelligent information process as diagnosis with less value-judgment will be treated in an effective way by artificial intelligence or knowledge engineering. However, among intelligent information process, the value-judgment will remain hard to treat to the last. In many practical problems such as portfolio problems, it is very important to get a solution reflecting faithfully the value-judgment of customers. In order to make decisions in a flexible manner for the multiplicity of value-judgment and complex changes of environment of decision making, the co-operative system of man and computers are very attractive: above all, interactive multi-objective programming methods seem promising.

Among several interactive multi-objective programming techniques, the aspiration level approach has been applied to several kinds of real problems, because

i. it does not require any consistency of judgment of DM,

ii. it reflects the value of DM very well,

iii. it is easy to implement.

In particular, the point (i) is very important, because DM tends to change his attitude even during the decision making process. This implies that the aspiration level approach such as the satisficing trade-off method can work well not only with the multiplicity of value judgment of DMs but also the dynamics of value judgment of them.

References

[1] Borde, J. and Crouzeix, J.P. (1987): Convergence of a Dinkelbach-type Algorithm in Generalized Fractional Programming, *Zeitscrift fur Operations Research* 31, 31-54.

[2] Charnes, A. and Cooper, W. (1961): *Management Models and Industrial Applications of Linear Programming, Vol. 1*, John Wiley, New York

[3] Eschenauer, H.A., Koski, J. and Osyczka, A. (1990): *Multicriteria Design Optimization*, Springer

[4] Ferland, A. and Potvin, J. (1985): Generalized Fractional Programming: Algorithms and Numerical Experimentation, *European J. Operational Research* 20, 92-101.

[5] Furukawa, K., Inoue, K. Nakayama, H. and Ishido, K. (1986): Management of Erection for Cable Stayed Bridge using Satisficing Trade-off Method, *Journal of the Japan Soc. of Civil Engineers*, 374/I-6, 495-502. (in Japanese)

[6] Grauer, M., Lewandowski, A. and Wierzbicki A.P. (1984): DIDASS Theory, Implementation and Experiences, in M. Grauer and A.P. Wierzbicki (eds.) *Interactive Decision Analysis, Proceedings of an International Workshop on Interactive Decision Analysis and Interpretative Computer Intelligence*, Springer: 22-30.

[7] Kamenoi, T., Sugimoto, H. and Nakayama, H. (1992): Studies on Revised Satisficing Trade-off Method for Structure Design, *Journal of the Japan Soc. of Civil Engineers*, **441/I-18**, 117-126 (in Japanese)

[8] Konno, H. nad Inori, M. (1987): Bond Portfolio Optimization by Bilinear Fractional Programming, *Research Paper IKSS 87-4, Institute of Human and Social Sciences*, Tokyo Institute of Technology

[9] Korhonen, P. and Wallenius, J. (1988): A Pareto Race, *Naval Research Logistics*, 35, 615-623

[10] Korhonen, P. (1987): VIG-A Visual Interactive Support System for Multiple Criteria Decision Making, *Belgian J. Oper. Res. Stat. Comp. Sci.* 27, 3-15

[11] Mitani, K. and Nakayama, H. (1994): Feed Formulation for Live Stock using Interactive Multi-objective Programming Techniques, *Agricultural Systems* (to appear in Japanese).

[12] Nakayama, H. (1984): Proposal of Satisficing Trade-off Method for Multi- objective Programming, *Transact. SICE*, 20: 29-35 (in Japanese)

[13] Nakayama, H. (1989): An Interactive Support System for Bond Trading, in A. G. Lockett and G. Islei (eds.) *Improving Decision Making in Organizations*, Springer, 325-333

[14] Nakayama, H. (1991-a): Satisficing Trade-off Method for Problems with Multiple Linear Fractional Objectives and its Applications, in A. Lewandowski and V. Volkovich (eds.) *Multiobjective Problems of Mahtematical Programming*, Springer, 42-50.

[15] Nakayama, H. (1991-b): Trade-off Analysis Based Upon Parametric Optimization, in P.Korhonen, A. Lewandowski and J. Wallenius (eds.), *Multiple Criteria Decision Support*, Springer, 42-52

[16] Nakayama, H. (1992): Trade-off Analysis using Parametric Optimization Techniques, *European J. of Operational Research*, **60**, 87-98

[17] Nakayama, H., Mitani, K. and Yoshida, F. (1992): A Support System for Multi-objective Feed Formulation, *Proc. of Software Conference*, Osaka, 17-20, (in Japanese).

[18] Nakayama, H., Nomura, J., Sawada, K. and Nakajima, R. (1986): An Application of Satisficing Trade-off Method to a Blending Problem of Industrial Materials, in G. Fandel et al. (Eds.), *Large Scale Modelling and Interactive Decision Analysis*. Springer, 303-313.

[19] Nakayama, H. and Sawaragi, Y. (1984): Satisficing Trade-off Method for Interactive Multiobjective Programming Methods, in M. Grauer and A.P. Wierzbicki (eds.) *Interactive Decision Analysis, Proceedings of an International Workshop on Interactive Decision Analysis and Interpretative Computer Intelligence*, Springer: 113-122.

[20] Nakayama, H. and Tanino, T. (1994): *Multi-objective Programming-Theory and Applications*, Society of Instrument and Control Engineering, (in Japanese)

[21] Sawaragi, Y., Nakayama, H. and Tanino, T. (1985): *Theory of Multiobjective Optimization*, Academic Press

[22] Ueno, N., Nakagawa, Y., Tokuyama, H., Nakayama, H. and Tamura, H., (1990): A Multi-objective Planning for String Selection in Steel Manufactureing, *Communications of the Operations Research Society of Japan*, **35**, 656-661 (in Japanese)

[23] Wierzbicki, A.P., (1981): A Mathematical Basis fro Satisficing Decision Making, in J.N. Morse (ed.), *Organizations: Multiple Agents with Multiple Criteria*, Springer-Verlag, 465-485

[24] Wierzbicki, A.P., (1986): On the Completeness and Constructiveness of Parametric Characterizations to Vector Optimization Problems, *OR Spektrum*, 8, 73-87

WHEN IS A STACKELBERG EQUILIBRIUM PARETO OPTIMUM? *

ATHANASIOS MIGDALAS samig@math.liu.se
Division of Optimization, Department of Mathematics
Linköping Institute of Technology, S - 581 83 Linköping, Sweden

Abstract. Bi-level optimization problems arise in hierarchical decision making, where players of different ranks are involved. The situation is described by the so-called Stackelberg game. The players of lower rank, called followers, react to decisions made by the first rank player, also called the leader. Situations similar to this arise for instance in mixed economies, land-use, traffic signal setting and in the particularly well-known network design problem. Previously, solution techniques based on the replacement of the bi-level problem by a bi-criteria one were proposed. However, it was shown subsequently, by the means of counter examples, that the bi-level problem is not equivalent to the bi-criteria formulation. In this note we demonstrate that the bi-criteria approach is not necessarily obsolete. We derive conditions under which we show that a Stackelberg equilibrium is indeed a Pareto optimum.

Keywords: Multi-objective programming; Bi-level programming; Stackelberg equilibrium; Pareto optimum; Network Design.

1. Introduction

The bi-level programming problem describes a hirerchical system that is composed of two levels of decision makers. The higher level decision maker, known as *leader*, controls the decision variables $y \in Y$, while the lower level decision maker, known as *follower*, controls the decision variables $x \in X$. The interaction between the two levels is modelled in their respective *loss functions* $\varphi(x, y)$ and $f(x, y)$. The leader and the follower play a *Stackelberg duopoly game* [16], [1]. The idea of the game is as follows: The first player, the leader, chooses $y \in Y$ to minimize the loss function $\varphi(x, y)$, while the second player, the follower reacts to leader's decision by selecting a strategy $x \in X$ that minimizes his loss function $f(x, y)$, in full knowledge of the leader's decision. Thus, the follower's decision depends upon the leader's decision, i.e. $x = x(y)$, and the leader is in full knoweledge of this. Consequently, we have the following definition:

Definition. If there exists a mapping $x : Y \to X$ such that for any fixed $y \in Y$,

$$f(x(y), y) \leq f(x, y), \quad \forall x \in X, \tag{1}$$

* I thank Dr. D.T. Luc for the discussions. This work has been partially supported by the Swedish Communication Research Board (KFB).

P.M. Pardalos et al. (eds.), Advances in Multicriteria Analysis, 175–181.

and if there exists $y^\star \in Y$ such that

$$\varphi(x(y^\star), y^\star) \;\leq\; \varphi(x(y), y), \;\; \forall y \in Y, \tag{2}$$

then the pair $(x^\star, y^\star)$, where $x^\star = x(y^\star)$, is called a *Stackelberg equilibrium* with the first player as leader and the second player as follower.

Bi-level programming models are derived from the conditions (1) -(2) of the definition in a most natural way. The graf $G = \{(x, y) \in X \times Y : x = x(y)\}$ of the mapping $x : Y \to Y$ represents the collection of strategy pairs in $X \times Y$ according to which the follower reacts to every strategy $y \in Y$ that the leader chooses. The follower is a *rational player* if he/she plays according to the set G. In the Stackelberg game, a basic assumption is that the follower is a rational player. According to the definition, the Stackelberg equilibrium prescribes an optimal strategy for the leader if the follower reacts by playing optimally, whenever the leader announce his move first. Mathematically we have the following results:

Definition. For given decision $y \in Y$ of the leader, the *rational reaction set* of the follower is

$$\begin{aligned} R_f(y) \;&=\; \{x(y) \in X : (x(y), y) \in X \times Y \;\text{ and} \\ &\qquad f(x(y), y) \leq f(x, y), \; \forall x \in X\} \end{aligned} \tag{3}$$

Definition. The leader's *set of admissible strategies* is

$$L_\varphi \;=\; \{y \in Y : \exists x \in R_f(y) \text{ such that } (x, y) \in X \times Y\} \tag{4}$$

Definition. A point $(x, y) \in X \times Y$ is *Stackelberg feasible* if $y \in L_\varphi$ and $x \in R_f(y)$.

THEOREM 1 *A Stackelberg feasible point $(x^\star, y^\star)$ is a Stackelberg equilibrium (with the first player as the leader) if $\varphi(x^\star, y^\star) \leq \varphi(x, y)$ for all $(x, y) \in X \times Y$.[15]*

Existence of Stackelberg equilibrium is guaranteed in the following case:

THEOREM 2 *If Y and X are compact sets in R^m and R^n respectively, and if φ and f are real-valued continuous functions on $X \times Y$, then a Stackelber equilibrium (with either player as leader) exists.[15]*

Consider the following bi-level program:

$$\min_{y \in Y} \quad \varphi(x(y), y) \tag{5}$$

$$\text{where } x(y) \;=\; \arg\min_{x \in X} f(x, y) \tag{6}$$

We assume here that the sets $X \subseteq R^n$ and $Y \subseteq R^m$ are convex and compact, that f and φ are continuous on $X \times Y$, and that (6) has a unique solution $x(y) \in X$ for all $y \in Y$. Thus, by Theorem 1.1, a Stackelberg equilibrium exists. Moreover, the mapping $x : Y \to Y$ is single-valued. These are not entirely unrealistic assumptions. The traffic network design problem [11], [13], [14], [5], [17], for instance, fulfill strong assumptions, i.e. the function f is strictly convex in $x \in X$ for all $y \in Y$. Moreover, the uniqueness of the optimal solution to the second level program is a standard assumption in the literature, even in the linear case [3], [20].

The range of applications modelled by (5-6) is broad. The Stackelberg games with the first player as leader are of particular practical interest when the information pattern is biased in the sense that the leader knows the loss functions of both players, while the follower is not informed about the leader's function. The follower is informed however about the leader's strategy and takes it into account when computing his own strategy. Therefore, of particular interest in the range of policy problems to which Stackelberg game is applicable are certain hierarchical decision-making systems in mixed economies where policy makers at the top level influence the decisions of private individuals and companies. In order to reduce a country's dependence on imported energy resources, for instance, a government can impose retail sales tax, import quotas and rationing. The energy consumption of individuals and companies will consequently adjust in accordance with respect to the resulting availability and prices. This will in turn affect import levels, the general price level and government revenues.

With respect to land-use policies [7], a government can decide to invest in order to provide transportation and communication infrastructures. Individual farmers and traders can then make their own decisions about production and marketing. Society depends heavily upon the mobility provided by road networks in order to make it possible for its members to participate in essential activities such as production, consumption and communication. Commodities are also transported from place to place because of the economic necessities of production and consumption and due to pursuit of economic advantages and gain. Deterioration of the conditions of the transportation system, especially in urban and metropolitan areas, has environmental, economic, health and safety impacts that are well-known. In order to reduce the negative effects and meet new demands, society invests huge amounts of money in the improvement and expansion of the existing road network. The network design problem [11], [13], [14] adresses some of these aspects.

Problems of type (5)–(6) arise also in traffic signal settings and urban network control [9], [8] as combined assignment and signal optimization. In analogy with the network design, the first level (5) concerns the decision of the network manager (the leader) who establishes the control system to be adopted on the basis of bublic interests. The second level (6) concerns the behaviour of the network users in the choice of their route. According to Wardrop's first principle [21], the individual trip maker is assumed to minimize only his own travel costs. The manager cannot prescribe the route for each individual, rather he predicts the user flow pattern by solving (6). The flow distribution must guarantee the minimization of user cost for each assigned regularization, i.e. user optimum. The manager (5) optimizes a function, expressing the collective cost, of control parameters and predicted flows according to Wardrop's second principle [21], i.e. system optimum.

We address the question whether the bi-level optimization problem can be solved as an equivalent bi-criteria optimization problem. Such an approach has been taken previously by [3], [20] for the linear case and subsequently by [11] in the case of the traffic network design problem which is a nonlinear problem. However, the approach was demonstrated inadequate in [?], [22], [14] with the means of counter examples. Concerning the facts that the general linear bi-level optimization problem is a global optimization problem [18], [19], [2] and NP-hard [10], [22], [6], the conclusions in [22], [14] should not surprise.

In this note we investigate conditions under which equivalence of the bi-level and bi-criteria optimization problems can be demonstrated.

2. Equivalence of bi-level and bi-criteria problems

Consider the bi-criteria problem

$$\min_{x\in X,\, y\in Y} \{\varphi(x,y), f(x,y)\} \tag{7}$$

and assume that the following assumptions are satisfied by (5)-(6):

ASSUMPTION 1 : $\forall y \in Y$, $x(y)$ is a unique solution of (6).

ASSUMPTION 2 : $f(x,y) \leq f(\hat{x},\hat{y}) \Rightarrow \varphi(x,y) < \varphi(\hat{x},\hat{y}),\;\; x,\hat{x} \in X, y,\hat{y} \in Y.$

For the vector problem (7), we recall (see e.g. [12] and [1]) the following definition:

Definition. The point $(x^\star, y^\star) \in X \times Y$ is an *efficient solution* or *Pareto optimum* of (7) if and only if there is no point $(x_o, y_o) \in X \times Y$ such that both

$$f(x_o,y_o) \;\leq\; f(x^\star,y^\star) \tag{8}$$

and

$$\varphi(x_o,y_o) \;\leq\; \varphi(x^\star,y^\star)\,, \tag{9}$$

where at least one inequality is strict.

Under Assumptions 1 and 2, we can prove the following result:

THEOREM 3 *If $(x(y^\star), y^\star)$ is a solution of (5)-(6), then it is also an efficient solution of the vector problem (7).*

Proof: Suppose to the contrary that $(x(y^\star), y^\star)$ is not a solution of the vector problem. Then, there exists some point $(x_o, y_o) \in X \times Y$ such that

$$\varphi(x_o,y_o) \;\leq\; \varphi(x(y^\star),y^\star) \tag{10}$$

$$f(x_o,y_o) \;\leq\; f(x(y^\star),y^\star) \tag{11}$$

and at least one of the inequalities is strict.

For fixed $y = y_o$, let $x(y_o)$ be the unique solution of (6). There are two possible cases:

Case I : $x_o = x(y_o)$. Then $\varphi(x(y_o), y_o) \leq \varphi(x(y^\star), y^\star)$, and due to the optimality of $(x(y^\star), y^\star)$, $\varphi(x(y_o), y_o) = \varphi(x(y^\star), y^\star)$. Thus, (10) is an equality. And, by Assumption 2, (11) is also an equality. A contradiction.

Case II : $x_o \neq x(y_o)$. Then $f(x(y_o), y_o) < f(x_o, y_o)$. And, by Assumption 2, $\varphi(x(y_o), y_o) < \varphi(x_o, y_o)$, which gives $\varphi(x(y_o), y_o) < \varphi(x(y^\star), y^\star)$ due to (10). Hence, contradicting the fact that $(x(y^\star), y^\star)$ is a solution of (5)-(6).

■

Conversely:

THEOREM 4 *If $(x^\star, y^\star)$ is an efficient point of (7) and $x^\star$ is a solution to (6) for fixed $y = y^\star$, then $y^\star$ is a solution to (5).*

Proof: By contradiction, if $y^\star$ is not a solution to (5), there exists $y_o \in Y$ and the corresponding solution $x(y_o) \in X$ to (6) so that

$$\varphi(x(y_o), y_o) < \varphi(x^\star, y^\star) \tag{12}$$

Since $(x^\star, y^\star)$ is an efficient point, and in view of (12), we must have $f(x(y_o), y_o) > f(x^\star, y^\star)$. By Assumption 2, the latter implies $\varphi(x(y_o), y_o) > \varphi(x^\star, y^\star)$, which contradicts (12). ■

3. Weakening the assumptions

A result similar to that of Theorem 3 can still be proved if Assumption 2 is weakened as follows:

ASSUMPTION 3 : $f(x, y) \leq f(\hat{x}, y) \Rightarrow \varphi(x, y) < \varphi(\hat{x}, y), \quad x, \hat{x} \in X, y \in Y.$

We recall from [1] and [12] the following definition:

Definition. The point $(x^\star, y^\star) \in X \times Y$ is said to be a *weak efficient solution* or *a weak Pareto optimum* of (7) if there is no point $(x, y) \in X \times Y$ such that both

$$\varphi(x, y) \; < \; \varphi(x^\star, y^\star) \tag{13}$$

and

$$f(x, y) \; < \; f(x^\star, y^\star) \; . \tag{14}$$

We can now prove the following:

THEOREM 5 *If $y^\star$ solves (5), then $(x(y^\star), y^\star)$ is a weak efficient solution to (7).*

Proof: By contradiction, if $(x(y^\star), y^\star)$ is not a weak efficient point, then there exists $(x_o, y_o) \in X \times Y$ such that

$$\varphi(x_o, y_o) \; < \; \varphi(x(y^\star), y^\star) \tag{15}$$
$$f(x_o, y_o) \; < \; f(x(y^\star), y^\star) \tag{16}$$

Let $x(y_o)$ be a solution of (6) for fixed $y = y_o$. Then $f(x(y_o), y_o) \leq f(x_o, y_o)$. By Assumption 3, $f(x(y_o), y_o) \leq f(x_o, y_o) \Rightarrow \varphi(x(y_o), y_o) < \varphi(x_o, y_o)$. Contradicting, due to (15), the fact that $y^\star$ solves (5). ■

3.1. Conclusions

We have demonstrated that under certain conditions a Stackelberg equilibrium is also a Pareto optimum and conversely. In view of this result, the bi-criteria approach taken by, for instance, [11] for the network design problem can be, to some extend, justified. However, further research on this subject is required. In particular, the further weakening of Assumption 2 is necessary in order to enlarge the problem class for which a Stackelberg equilibrium is also a Pareto optimum. An important question is whether such a weakening can still guarantee results similar to Theorem 4.

References

1. J.P. Aubin (1979) Mathematical Methods of Game and Economic Theory. North-Holland, Amsterdam.
2. F.A. Al-Khayal, R. Horst and P.M. Pardalos (1992) Global Optimization of Concave Functions subject to Separable Quadratic Constraints - An Application to Bi-Level Programming, Annals of Operations Research, vol. 34, pp. 125-147
3. J. F. Bard (1983) An Efficient Point Algorithm for a Linear Two-Stage Optimization Problem, Operations Research, vol. 31, pp. 670-684
4. O. Ben-Ayed and C.E. Blair (1990) Computational Difficulties of Bilevel Linear Programming, Operations Research, vol. 38, pp. 556-560
5. O. Ben-Ayed, C.E. Blair, D.E. Boyce and L.J. LeBlanc (1992) Construction of Real-World Bilevel Linear Programming Model of the Highway Network Design Problem, Annals of Operations Research, vol. 34, pp. 219-254
6. C. Blair (1992) The Computational Complexity of Multi-Level Linear Programs, Annals of Operations Research, vol. 34, pp. 13-19
7. W.R. Blundet and J.A. Black (1984) The Land-Use / Transport System. 2nd edition. Pergamon Press, Australia.
8. G.E. Cantarella and A. Sforza (1987) Methods for Equilibrium Network Traffic Signal Setting, in "Flow Control of Congested Networks", A.R. Odoni et al. (ed.s), NATO ASI Series, vol. F38, Springer-Verlag, Berlin, pp. 69 - 89
9. G. Improta (1987) Mathematical Programming Methods for Urban Network Control, in "Flow Control of Congested Networks", A.R. Odoni et al. (ed.s), NATO ASI Series, vol. F38, Springer-Verlag, Berlin, pp. 35 - 68
10. R.G. Jeroslow (1985) The Polynomial Hierarchy and Simple Model for Competitive Analysis, Mathematical Programming, vol. 32, pp. 131-153
11. L.J. Leblanc and D.E. Boyce (1986) A Bilevel Programming Algorithm for Exact Solution of the Network Design Problem with User Optimal Flows, Transportations Research, vol. 20B, pp. 259-265

12. D.T. Luc (1989) Theory of Vector Optimization. Lecture Notes in Economics and Mathematical Systems 319, Springer-Verlag, Berlin.
13. P. Marcotte (1983) Network optimization with continuous control parameters, Transportation Science, vol. 17, pp. 181-197
14. P. Marcotte (1988) A Note on the Bilevel Programming Algorithm by Leblanc and Boyce, Transportations Research, vol. 22B, pp. 233-237
15. M. Simaan and J.B. Cruz, Jr (1973) On the Stackelberg Strategy in Nonzero-Sum Games, JOTA, vol 11, pp. 533-555
16. H. von Stackelberg (1952) The Theory of the Market Economy. Oxford University Press.
17. S. Suh and T. Kim (1992) Solving Nonlinear Bilevel Programming of the Equilibrium Network Design Problem - A Comparative Review, Annals of Operations Research, vol. 34, pp. 203-218
18. H. Tuy, A. Migdalas and P. Värbrand (1993) A Global Optimization Approach for the Linear Two-Level Program, Journal of Global Optimization, vol. 13, pp. 1-23
19. H. Tuy, A. Migdalas and P. Värbrand (1994) A Quasiconcave Minimization Method for Solving Linear Two-Level Programs, Journal of Global Optimization, vol. 4, pp. 243-263
20. G. Unlu (1987) A Linear Bilevel Programming Algorithm Based on Bicreteria Programming, Computers and Operations research, vol. 14, pp. 173-179
21. J.G. Wardrop (1952) Some Theoretical Aspects of Road Traffic Research, Proceedings of the Institution of Civil Engineering, pp. 325-362
22. U.-P. Wen and S.-T. Hsu (1989) A Note on a Linear Bilevel Programming Algorithm Based on Bicreteria Programming, Computers and Operations Research, vol. 16, pp. 79-83

V. REAL-WORLD APPLICATIONS

PREFERENCE DISAGGREGATION ANALYSIS IN AGRICULTURAL PRODUCT CONSUMER BEHAVIOUR

Y. SISKOS, E. GRIGOROUDIS and N.F. MATSATSINIS
Technical University of Crete
Decision Support Systems Laboratory
73133 Chania, Greece

G. BAOURAKIS
Mediterranean Agronomic Institute of Chania
P.O. Box 85
73100 Chania, Greece

ABSTRACT. This article presents an accomplishment of the disaggregation analysis on the agricultural products consumer preferences. This is an alternative methodology to the product design problem in the area of agricultural marketing. This methodology applies a preference disaggregation analysis for each consumer separately in order to specify new brand ideas with the maximum possible ideas. Techniques related to market penetration strategy formulation are also utilised in the new product development. Finally, some characteristic results are presented through the implementation of the disaggregation model based on preferential data which are collected in the French olive oil market

Keywords: *preference disaggregation analysis, consumer behaviour, agricultural marketing*

1. Introduction

The preference disaggregation paradigm in multicriteria analysis was firstly initiated by Jacquet-Lagrèze at the end of the 70's aiming at defining, under certain assumptions, the multicriteria model that is hidden behind the overall preferences of a decision-maker. In the case of UTA method of the aforementioned authors (see section 2), the disaggregated overall preference takes the form of a ranking of some stimuli, that are called in this paper reference actions and the multicriteria model that is to be evaluated is an additive value function; the latter is achieved through the employment of linear programming techniques in a way that the multicriteria value function will give the same or at least similar ranking to the one the decision-maker has externalised.

P.M. Pardalos et al. (eds.), Advances in Multicriteria Analysis, 185–202.

The disaggregation analysis presumes the existence of overall preferences, which is not always feasible. On the other side it offers the possibility of an a posteriori quantification of the qualitative evaluations based on criteria in which case modelling is not feasible unless qualitative evaluation scales are employed.

This paper consist of an accomplishment of the disaggregation analysis on the agricultural products consumer preferences and more specifically of olive oil product. The preferential data (the overall ranking of olive oil brand and the evaluations on separate criteria) collected through the conduct of surveys that involve a certain sample of consumers. It is an alternative to the product design problem in the area of agricultural marketing approach (Baourakis, 1994; Ritson and Fearne, 1995).

Section 2 consists of a brief description of the UTASTAR model, which is utilised in this paper for the disaggregation of the consumer preferences as well as of techniques related to market penetration strategy formulation. Section 3 presents some statistics related to the European olive oil market as well as the survey of preference collection in the Paris olive oil market; the part of the questionnaire used for this purpose (translated in English) is in the Appendix. Finally, section 4 outlines some characteristic results from this analysis and its economic implications.

2. Preference disaggregation of consumer choices

Basic Ideas

A consumer choice policy can be externalised by means of a set of reference products (brands) which the consumer, either has or can rank through simple questionnaires, familiar decision making situations, and so on.

The disaggregation analysis problem concerned here is how to estimate for a given consumer a multiattribute utility or value function which should be as consistent as possible with the consumer ranking of these reference products.

The UTASTAR model (Siskos and Yannacopoulos, 1985), an improved version of the multicriteria method UTA of Jacquet-Lagrèze and Siskos (1982), is applied on the consumer data preference set. This is an ordinal regression method allowing the use of qualitative criteria. During the survey each consumer is asked to express his/her estimation for a set of products $A = \{a_1, a_2, \ldots, a_m\}$ which are examined on the basis of certain criteria. Following this, all products are ranked by the purchaser according to the order which he/she would purchase them on the assumption that the product previously selected was not available. This estimates the consumer's utility function which has the following additive form:

$$u(\underline{g}) = p_1 u_1(g_1) + p_2 u_2(g_2) + \ldots + p_n u_n(g_n) \tag{1}$$

$$\text{s.t.} \qquad u_i(g_{i*}) = 0, \;\; u_i(g_i^*) = 1 \;\forall\; i = 1,2,\ldots,n, \quad \sum_{i=1}^{n} p_i = 1$$

where $\underline{g} = (g_1, g_2, \ldots, g_n)$ is a vector of the utilised criteria, g_{i*} and g_i^* the worst and the best level of criterion i respectively, $u_i(g_i)$ is the marginal utility of criterion g_i normalised between 0 and 1, $u(\underline{g})$ is the global utility of $\underline{g}$ and p_i is the weight of the i-th criterion. The sum of weights is one. The utility function is as consistent as possible with the observed selections on known subjective preferences. The revised UTA method (UTASTAR) estimates an equivalent unweighted utility function, as follows below.

The UTASTAR algorithm

Given a subjective consumer ranking ($\succ$, $\sim$) on a set of reference products $A_R = \{a_1, a_2, \ldots, a_k\}$, where they are rearranged so that $a_j \succ a_{j+1}$ (a_j preferred to a_{j+1}), or $a_j \sim a_{j+1}$ (a_j indifferent to a_{j+1}), for every $j = 1, 2, \ldots, k-1$, the UTA algorithm procedure runs in four steps:

Step 1

Expresses the global utilities of reference products $u[\underline{g}(a_j)]$, $j = 1, 2, \ldots, k$, first in terms of marginal utilities $u_i(g_i^1)$, then in terms of variables:

$$w_{ir} = u_i(g_i^{r+1}) - u_i(g_i^r) \geq 0, \; i = 1, 2, \ldots, n; \;\; r = 1, 2, \ldots, a_i-1 \tag{2}$$

by means of the relations

$$u_i(g_i^r) = 0 \text{ and } u_i(g_i^r) = \sum_{t=1}^{r-1} w_{it}, \; \forall \text{ i and } r > 1 \tag{3}$$

Step 2

Introduces two error functions σ^+ and σ^- by writing for each pair of consecutive products in the ranking, the analytic expressions:

$$\Delta(a_j, a_{j+1}) = u[\underline{g}(a_j)] + \sigma^+(a_j) - \sigma^-(a_j) - u[\underline{g}(a_{j+1})] - \sigma^+(a_{j+1}) + \sigma^-(a_{j+1}) \tag{4}$$

Step 3

Solves the linear program:

$$\text{Minimize } F = \sum_{j=1}^{k} [\sigma^+(a_j) + \sigma^-(a_j)] \tag{5}$$

under the set of constraints:

for $j = 1, 2, ..., k - 1$:

$$\Delta(a_j, a_{j+1}) \geq \delta \text{ if } a_j \succ a_{j+1} \tag{6}$$

$$\Delta(a_j, a_{j+1}) = 0 \text{ if } a_j \sim a_{j+1} \tag{7}$$

$$\sum_i \sum_r w_{ir} = 1$$

$$w_{ir} \geq 0, \quad i = 1, 2, ..., n; \; r = 1, 2, ..., \alpha_i - 1 \tag{8}$$

$$\sigma^+(a_j) \geq 0, \; \sigma^-(a_j) \geq 0, \; j = 1, 2, ..., k \tag{9}$$

δ being a small positive number.

Step 4

Test the existence of multiple optimal or near optimal solutions of the linear program (4) - (8) (stability analysis); in case of non uniqueness, find the mean additive utility of those (near) optimal solutions which maximize the objective functions $p_i = u_i(g_i^*) = \sum_r w_{ir}$ for all i = 1, 2, ..., n on the polyhedron (6) - (9) bounded by the new constraint:

$$\sum_{j=1}^{k} [\sigma^+(a_j) + \sigma^-(a_j)] \leq F^* + \varepsilon \tag{10}$$

F^* being the optimal value of the linear program in step 3 and ε a very small positive number.

Product development research methodology

This section outlines a new product development procedure which is based on a large scale consumer study and incorporates all the necessary steps for an effective marketing research: definition of the problem and research objectives, development of the research plan, collection of data and information, processing and analysis utilising multidimensional and multicriteria models and finally presentation and elaboration of results (cf. figure 1)

A first stage in applying the methodology, includes the examination and selection of competing products, a preliminary market research on supply and demand of the brands through secondary sources of data, the questionnaire preparation, sampling, determination of contact method, conduct of a pilot study, final modification of the questionnaire and survey completion.

The proposed methodology operates by means of an integrated decision support system (DSS) (Siskos and Matsatsinis, 1993). Detail information about the models used are given below.

Determinant criteria and market shares estimation

Based on the values of the criteria weights $\underline{p}$ and of an important threshold, 10% for example, it is possible to underline the determinant criteria, in other words these that play the most important role, from the very beginning in the consumer level and at second stage for all consumers, through the estimation of the percentage of consumers for whose each one criterion is appeared as determinant.

Market shares forecasting

The market shares estimation is based on the utility functions estimated by the UTASTAR model for a set of consumers: J = {1, 2, ..., k}. This is accomplished by evaluating a product's purchasing probabilities for each single consumer, based on the formulae:

i) **Luce** (1959, 1977):

$$P(a_i; A) = u[\underline{g}(a_i)] / \sum_{l=1}^{m} u[\underline{g}(a_l)] \qquad i=1,2,...,m \qquad (11)$$

ii) **Lesourne** (1977):

$$P(a_i; A) = (u[\underline{g}(a_i)])^2 / \sum_{l=1}^{m} (u[\underline{g}(a_l)])^2 \qquad i=1,2,...,m \qquad (12)$$

iii) **Maximum Utility Formula**:

$$P(a_i; A) = \begin{cases} \dfrac{1}{R} & \text{when } u[\underline{g}(ai)] = \max\limits_{a\in A}\{u[\underline{g}(a)]\} \\ 0 & \text{otherwise} \end{cases} \qquad (13)$$

where $P(a_i;A)$ is the purchasing probability of the product a_i from the set A of available products, and R is the number of products providing the maximum utility.

Therefore for each consumer j, the probability of purchasing each market product is

$$[P_j(a_1), P_j(a_2),...,P_j(a_m)]^T, \qquad j=1,2,...,k \qquad (14)$$

The evaluation for each single product is retrieved

$$S_i = \sum_{j=1}^{k} P_j(a_i) \qquad i=1,2,...,m \qquad (15)$$

Finally, the market share for product a_i is estimated:

$$MA(a_i) = 100 * S_i \,/ \sum_{l=1}^{m} S_l \,\% \qquad i=1,2,\ldots,m \qquad (16)$$

Implementation

All the products examined in the conducted survey form the product base of the utilised DSS system which provides the flexibility of either inserting new products or discarding others. The system user has the opportunity of creating new partial reference markets. The data from the initial reference market are processed applying the three formulae of the previous section.

The results, compared with the existing product market shares, identify the formula which will be implemented. New market shares are generated by experts introducing multicriteria evaluations of the product under development. The new product utility is estimated according to the appropriate formula. Various market penetration scenarios for the new product are feasible through the modification of the multicriteria evaluations. The outcomes, based on the different strategies, are compared and the optimum market policy is selected for the firm (Fig. 1).

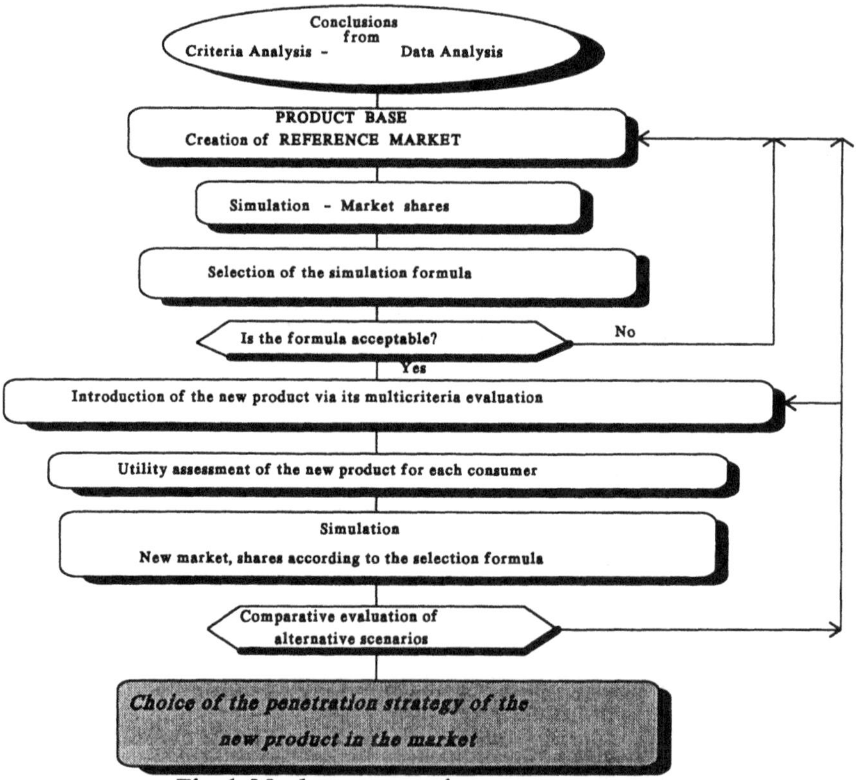

Fig. 1. Market penetration strategy process
Source: Siskos and Matsatsinis (1993)

3. The French olive oil market: Survey design

International Olive Oil Market

Olive oil production fluctuates from one crop year to another. This has an impact on prices, producer income and constant supply. Italy, Spain, and Greece are the main EU producers, followed by Portugal and France (Fig. 2).

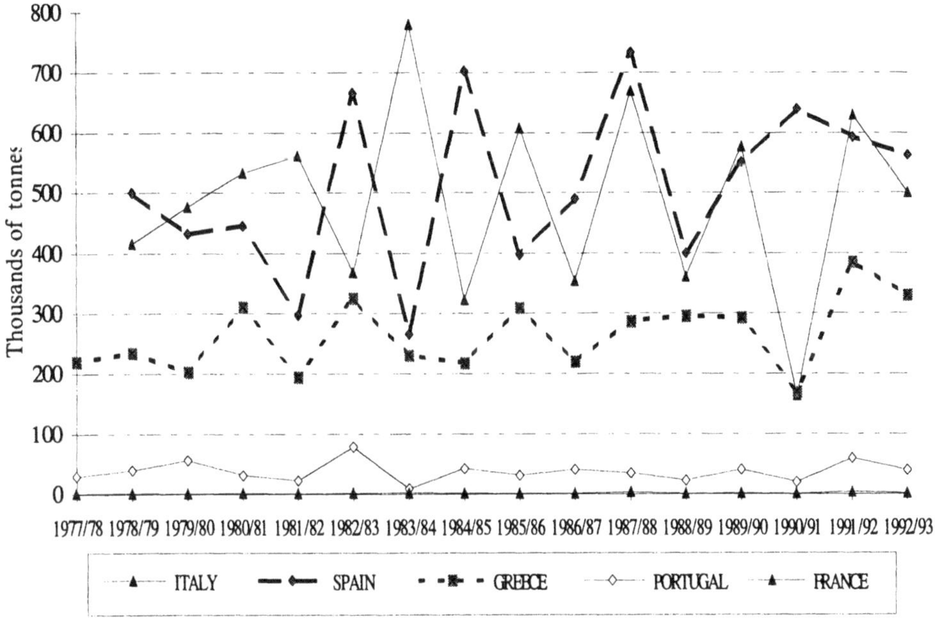

Fig. 2. Olive oil production in Italy, Spain, Greece, Portugal and France, 1977-1993 (thousands of tonnes). Source: IOOC (1991, 1992)

The decrease in Italian and Greek production in 1990 had an immediate impact on olive producer prices, while the sharp increase in the olive oil production of the same countries decreased the producer prices drastically. The increased price differential between olive oil and its close substitutes, the vegetable oils, has led to a rapid increase in the consumption of the latter.

On the demand side, world consumption is independent of the rate of production. EU olive oil consumption accounts for most of the world consumption. The traditional olive oil EU producer - states are the major consumers as well. Greece has had a stable consumption rate of 20 kg/person, in the 1977-93 period. Italian and Spanish consumption has varied with a slow increasing tendency while consumption in Portugal and France has been rising steadily (Fig. 3).

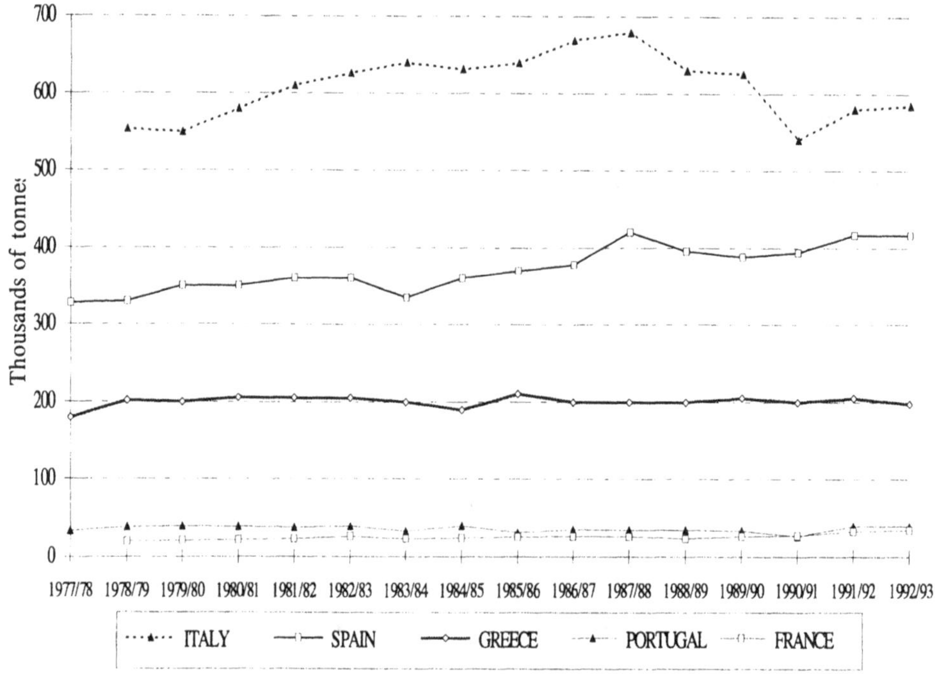

Fig. 3. Olive oil consumption in Italy, Spain, Greece, Portugal, France, 1977-1993 (thousands of tonnes). Source: IOOC (1991, 1992)

EU - olive oil economic policies

The olive oil market in the EU, in the period of 1977-1980 and 1981-1985 (EC of 10) experienced an overall shortage. Since the integration of Spain and Portugal into the EU (1985/86-1992/93), olive oil has been in surplus. Forecasting techniques (Box - Jenkins) applied on the available olive oil production and consumption data, revealed a persistence of the production surplus.

The European Community regulations for olive oil originate from the Common Market Organisation (CMO) for olive oil which was established in 1966 (CEE. 136/66) for a common market for fats and oils. The importance of these regulations is illustrated by the fluctuations in olive oil production from one crop year to another. Production decreased by 37.5% (1987-88) and 31% (1989-90), and increased by 35.6% (1988-89) and 50% (1990-91). The role of the olive oil product within EU has been upgraded since all three main producer countries Spain, Italy and Greece are members of the Community. Among the main objectives of EU policy are restriction of production and increase in consumption over its main vegetable substitutes.

The olive oil price intervention system is complex, and includes the producer target price, the intervention price, the representative market price, the threshold price and the utilisation of a variable levy for imports and a variable refund for exports. Compensatory payments are also part of this system, offering production and consumption aid. Production aid is available to all EU olive oil producers and is approximately the difference between the producer target price and the representative market price. Consumption aid is available to olive oil packaging firms, applicable to containers of 5 litres maximum capacity.

Analysis of consumer behaviour and market trends

Consumer behaviour analysis is accomplished utilising multicriteria preference analysis for each consumer separately in combination with data analysis methods and techniques concerning consumer profiles in general.

A consumer selection policy can be externalised by means of a set of reference products which he/she, the consumer, either has or can rank through simple questionnaires, familiar decision making situations, and so on.

The problem concerned here is how to estimate a multiattribute utility under certainty which should be as consistent as possible with the consumer ranking of these reference products. Results expected from the utilisation of the UTASTAR model are, examination of consumer behaviour determinant criteria and identification of consumer market segmentation.

Descriptive statistics and data analysis methods help to determine the profile of the consumer purchasing each of the examined products (Bouroche, 1977). The analysis has two directions: one toward the identification and determination of consumer general perceptions and attitudes and the other toward the examination of the existing interactions among consumers on different product criteria and the correlation among the attributes of the products. The data analysis methods used in this market research were principal components analysis and correspondence analysis.

Survey and First Results

A survey was conducted in Paris, France during the months of January- February, 1993.

Olive oil consumption in France is very low in comparison to the one in Italy, Spain, and Greece but quite significant with respect to the rest of European countries. The French olive oil production is mainly concentrated in the southern part and the existing gap between production and consumption level has been increasing during the last years.

Six olive oil brands were utilised: the Italian Carapelli, the Spanish Lerida, the Greek Kolymvari and the French olive oil brands Hediard, La Jarre d' Or and Puget.

Two market segments initially appear, the one consisting of the traditional consumers of the south and the one in the north, which is behaving similarly to the north European oil consumers.

All the participating brands were of half a litre volume, the first five, were extra virgin olive oil products, while La Jarre d' Or was a biological olive oil product.

Carapelli being a medium quality extra virgin olive oil seems to be competitive to the Cretan olive oil. Lerida also is one of the few Spanish branded products which are exported to France. La Jarre d'Or is available in health food stores while Hediard can be purchased only in Hediard stores, which provide products of high quality. Finally, Puget is available in almost every supermarket and is considered as the leader of the French olive oil market.

The questionnaire was designed in such a way in order to collect information of the sample's demographic characteristics, determine the oils generally used, frequency of olive oil use, type of use of the different oils and consumer perceptions with respect to attributes and factors such as health, taste, purity and pricing. Another section dealt with the size and type of packaging, brand loyalty, branding, colour, odour, taste, packaging and finally labelling. Through the last part of the questionnaire the aim was to identify afterwards which one of the six olive oil brands the consumer would purchase in a ranking order. The part of the questionnaire related to consumer preferences (marginal and overall) appears in Appendix.

Two hundred and five olive oil consumers were interviewed in their houses. Interviews lasted approximately 45 minutes and in each single case all the examined brands were evaluated. A number of districts of different socio-economic background were included in this survey.

Nearly all consumers utilise olive oil in general, 79.9% of them use sunflower oil while 50% use groundnut oil (Table 1).

The frequency of olive oil use classifies the consumers into two broad categories; the regular consumers (75.1%) and the occasional consumers (24.9%).

Table 2 shows that consumers mainly use olive oil for salads, hot dishes and cooking. In frying, sunflower and groundnut oils are used, while in cooking, sunflower, olive oil and to a lesser extent, margarine. For the preparation of mayonnaise and sauces sunflower and olive oil are used, while for salads and hot dishes the order of usage is reversed.

Oils	%
Olive oil	99.5
Groundnut oil	50.0
Sunflower oil	79.9
Soya oil	0.0
Corn oil	16.2
Margarine	36.3
Other oils	33.8

Frequency of olive oil use	%
Daily	22.1
3 - 4 times/week	26.0
1 - 2 times/week	27.0
Every 15 days	9.8
Every month	15.1

Table 1. Oils used generally and frequency of olive oil use.

	Frying (%)	Cooking (%)	Mayonnaise (%)	Sauces (%)	Salads (%)	Hot Dishes (%)
live	7.4	44.1	29.9	28.9	97.5	53.4
roundnut	31.4	19.6	22.1	18.1	9.8	15.7
unflower	39.7	48.0	48.0	39.7	27.9	35.8
orn	2.9	9.8	4.9	6.9	1.5	8.8
argarine	4.4	25.5	1.5	6.4	0.0	5.4
ther	11.3	12.3	1.5	7.4	10.8	9.3

Table 2. Frequency and type of oil use

4. Results and Recommendations

Determinant Criteria

Disaggregation analysis allows for the extraction of the criteria that appeared as determinant of the consumer preferences, which are those that the overall significance overcomes an importance threshold of 10%.

Among consumers, 80% seem to consider the product's image to be important, while a 77% considers the olive oil colour as a determinant criterion.

Criteria	Consumers (%)
Image	80
Colour	77
Packaging	70
Price	64
Taste	61
Odour	57

Table 3. Determinant criteria

Based on the criteria ranking it is feasible to segment the olive oil market through criteria or complex combinations of them. The criteria of price and colour are determinant for 49% of consumers (Table 4), while 37% of them perceive as important the image, the price and the packaging and they do not consider as determinant the criteria of colour, taste and olive oil odour (Table 5).

	Color (%)	Image (%)	Price (%)	Packaging (%)	Taste (%)	Odour (%)
Color	77	61	49	58	48	47
Image		80	53	54	50	46
Price			64	46	37	35
Packaging				70	43	44
Taste					61	41
Odour						57

Table 4. Market segmentation through criteria combinations

Color	Image	Price	Packaging	Taste	Odour	Consumers
+				+	+	34%
+	-	-	-	+	+	0%
+		+		+		29%
+	-	+	-	+	-	0%
		+	+			46%
-	-	+	+	-	-	0%
	+	+	+			37%
-	+	+	+	-	-	2%
+		+	+	+		22%
+	-	+	+	+	-	0%

Table 5. Market segmentation through complex criteria combinations

The decisive conclusion coming from this analysis is that all six criteria are segmented, generally, into three categories according to their importance. Olive oil image and colour are the most determinant criteria for the choice of a specific olive oil brand, while packaging and price consist a criteria group of medium importance to French consumer. Finally, olive oil taste and odour are the least determinant criteria for the consumer behaviour.

Estimation of market shares

The present market shares in the French olive oil market including the Kolymvari product are presented in Figure 4. It appears that the market leader by far is Puget (45%), followed by Carapelli (15%) and Hediard (16%). Kolymvari would benefit a serious market share reaching a 15%. This percentage will be mainly extracted from Puget and Carapelli.

Olive oil market forecasting

At this stage we will try to put the consumers' preferences and tendencies revealed in the previous analysis, in some order and, furthermore, design and develop the new Kolymvari olive oil product. This new product will be a modification of the extra virgin Kolymvari brand in order to better satisfy the consumer needs and adjust to the guidelines set by the French market. The utilised software for the qualitative marketing has the potentiality of market simulation through the conducted survey's data base.

The simulation methodology is based on each consumer utility functions which are estimated with the UTASTAR. By modifying the attributes-criteria of a product it can be evaluated for each individual consumer the utility of purchasing this new product. A new reestimation of the market shares is taking place then.

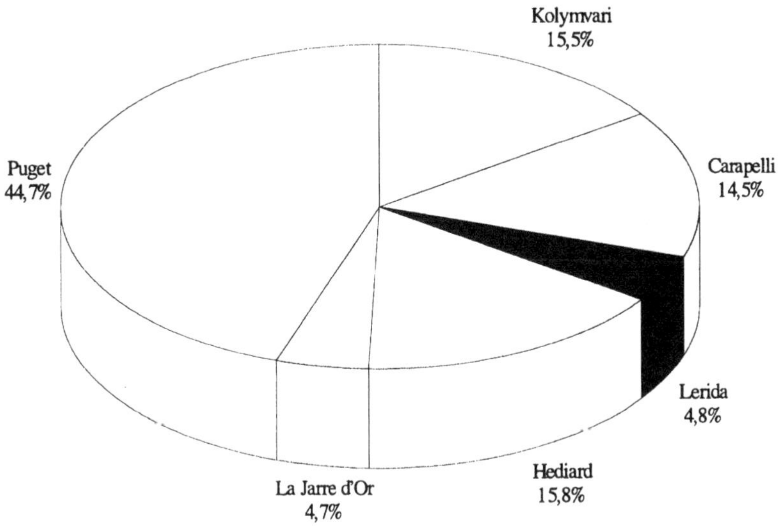

Fig. 4. Estimation of market shares in France (including Kolymvari)

The brands of La Jarre d' Or, Hediard and Lerida are not involved in these analyses since the consumers that purchase them are quite loyal. Therefore any possible change and improvement on the Kolymvari's attributes will not bring a significant impact on these three products' market shares.

The most significant criteria are price, packaging, colour and taste. This comes as a conclusion since they induce the greatest impact on the market shares. The odour and image of the product are less important criteria.

Potential market penetration strategies

The values of the criteria for the Kolymvari olive oil product are displayed in Table 6. The new product design will mainly be based on the improvements of the determinant criteria of colour, taste and packaging, while price is treated as an independent variable.

Influence		Packaging	
Bad	2.0	Bad	15.7
I do not know	61.8	Fair	47.1
Good	22.5	Good	32.8
Very good	13.7	Very good	4.4
Color		**Odour**	
Unnatural	16.7	Unnatural	37.3
Natural	70.1	Natural	46.1
Attractive	13.2	Pleasant	16.7
Taste			
Unnatural	25.5		
Natural	39.2		
Delicious	35.3		

Table 6. Distributional evaluations of the Kolymvari product

Table 7 presents three alternative market penetration strategies for the Kolymvari product in the French market.

Criteria	Strategy 1	Strategy 2	Strategy 3
Price	18-37F	18-37F	18-37F
Packaging	Fair	Good	Good
Taste	Natural-Delicious	Natural	Natural-Delicious
Odour	Natural	Natural	Natural
Colour	Natural	Natural	Natural
Image	Unaware-Good	Unaware-Good	Unaware-Good

Table 7. Alternative market penetration strategies for the Kolymvari product

An improvement on the taste is recommended in order for Kolymvari to approach the lighter taste of Lerida and Hediard. The present dark green colour of Kolymvari, while it is suitable for south France, it is not ideal for the greater area of Paris. Consumers have ranked the examined products from darker to lighter in the following order: Kolymvari - Lerida - Hediard - Carapelli. The new product's colour ought to look like the ones of Lerida and Hediard. The determination of the colour mainly depends on factors such as the type and quality of olive trees.

Finally a new packaging is required such as to approximate the evaluation of good. The packaging of Hediard and Carapelli has been rated as good. The required modifications are that:

- the label has to be in French (which was not the case for the utilised labelling of Kolymvari in this survey);
- the shape of the glass bottle has to be square with small diameter and long height and neck (Carapelli and Hediard have this type of shape);
- the container needs to open easily, pour and reseal properly ;
- a more intense promotion of the product's Cretan origin is required (both Carapelli and Lerida emphasize it);

These three strategies offer alternative solutions to the decision maker. The first is based on the natural attributes' improvement (colour, taste), the second on the packaging and the last on a combination of the previous two: improvement of packaging, colour and taste (Baourakis, 1994).

A comparison of these three different approaches (Fig. 5) indicates that the first two provide rather similar sales revenues while the third one appears as the optimum. This was expected since all three involved criteria were improved.

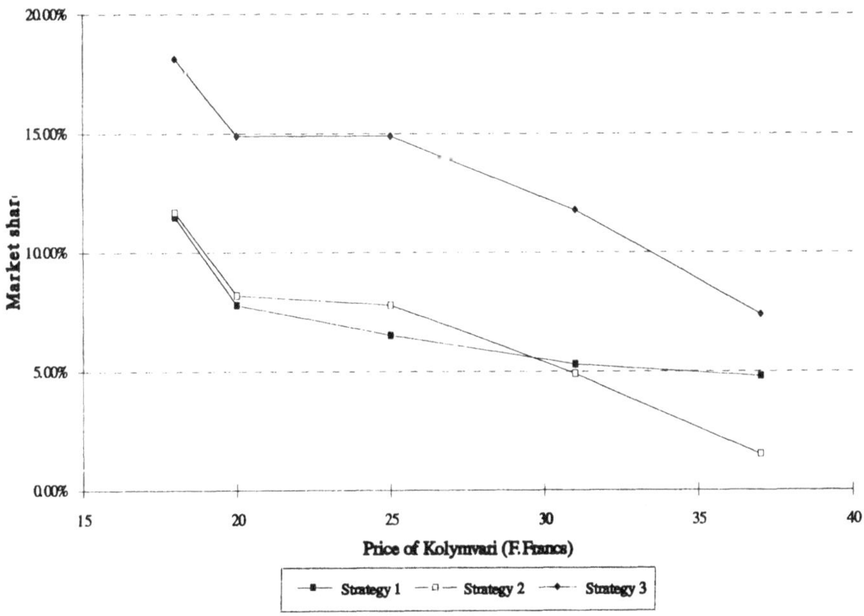

Fig. 5. Market shares (sales value) of Kolymvari for the three potential marketing strategies

Concerning the price of Kolymvari it can increase from 20FF to 25FF without a significant decrease of its market share. If it approaches 31FF then it is not a competitive product since Carapelli has the same price as well. On the contrary, if its price decreases to 18FF, that of Puget's, then its market share is further increased.

Recommendations

The French oil market is segmented into the north and south consumers. The southern part mainly consists of traditional olive oil consumers who mainly prefer a dark green colour. On the contrary, in the northern areas the use of olive oil is in very limited amounts and the main influence on the decision to use olive oil arises from the health benefits.

This survey took place in the greater area of Paris which can be considered as a representative sample of the country since both types of oil consumers are present. The olive oil market in Paris is not homogeneous and to a large extent the consumers' behaviour is determined by their origin.

The olive oil market in France and mainly in the northern part is of great interest since olive oil is recognised as a product of higher status. A percentage of 50-60% who presently purchase Puget or Carapelli would be willing to try the new Kolymvari product.

Price seems to be the most important determinant criterion. Despite that, if consumers are persuaded of a product's superior qualitative characteristics, they prefer it. This is the case of Hediard and its consumers are loyal to it as well. There is also a small market segment (5%) loyal to the biological olive oil La Jarre d' Or.

If we plan to introduce an expensive olive oil product to the French market then a great effort is required to assure potential buyers that this product is worth it. A great consistency is expected with respect to the taste, colour and odour of the product. Advertisement and promotion of the Cretan Green olive oil Kolymvari product with its superior health and organoleptic attributes as a first cold extraction product is highly recommended.

APPENDIX: PART OF QUESTIONNAIRE FOR CONSUMER PREFERENCE COLLECTION

What is your opinion about the following products?

PRODUCTS		CARAPELLI	LERIDA	KOLYMVARI	HEDIARD	LA JARRE d'OR	PUGET
ENVIRONMENTAL INFLUENCE (a)	Bad						
	I do not know						
	Good						
	Very Good						
COLOR (b)	Unnatural						
	Natural						
	Attractive						
ODOUR (c)	Unnatural						
	Natural						
	Pleasant						
TASTE (d)	Unnatural						
	Natural						
	Delicious						
PACKAGING (e)	Bad						
	Fair						
	Good						
	Very Good						

(a) According to what you know or have heard, what is your opinion of these brands?
(b) How would you characterise the color of these olive oils?
(c) How would you characterise the odour of these olive oils?
(d) How do you find the taste of these olive oils?
(e) How do you find the packaging of these olive oils?

RANKING
Which of the following extra virgin olive oils would you buy?
Repetitive question: If you could not find your oil of choice, which one would you buy?

CARAPELLI	LERIDA	KOLYMVARI	HEDIARD	LA JARRE D'OR	PUGET

References

Baourakis G. (1994). Agricultural product design methodology: The case of olive oil, Ph.D Thesis, Technical University of Crete.

Bouroche J.M. (1977). Analyse des Données en Marketing, Masson, Paris.

International Olive Oil Council (1991). Chronological Statistical Series, Madrid.

International Olive Oil Council (1992). International trade standard applying to olive oils and olive pomace oils, May 1992.

Jacquet - Lagrèze E. and J. Siskos (1982), Assessing a set of additive utility functions for multicriteria decision making : the UTA method, European Journal of Operational Research, 10 (2) 151-164.

Lesourne J. (1977), A theory of the individuals for economic analysis, North-Holland publishing company, 1, New York.

Luce R. (1959), Individual choice behaviour, John Wiley and Sons, New York.

Luce R. (1977), The choice axiom after twenty years, Journal of Mathematical Psychology, 15, 215-233.

Ritson C. and A. Fearne (1995). Forecasting agricultural commodity markets under policy constrains: A retrospective case study, in: J. Janssen, C. H. Skiadas and C. Zopounidis (eds.), Advances in Stochastic Modelling and Data Analysis, Kluwer Academic Pub.

Siskos J. and D. Yannacopoulos (1985), UTASTAR: An ordinal regression method for building additive value functions, Investigaçao Operacional, 5 (1), 39-53.

Siskos Y. and N. F. Matsatsinis (1993), A DSS for market analysis and new product design, Journal of Decision Systems, 2 (1).

AN APPLICATION OF THE UTA DISCRIMINANT MODEL FOR THE EVALUATION OF R & D PROJECTS

Eric Jacquet-Lagrèze
Euro-Décision, 36 rue des Etats généraux, 78000 Versailles, France

Abstract : *This paper presents a real world application of the UTA disaggregation method used to implement a formal multicriteria evaluation procedure in R & D projects management. The study includes the definition of the criteria, the choice of an aggregation procedure and the methodology used in order to assess the weights of the criteria. The formal multicriteria evaluation procedure is in use in the company since the study was performed.*

1. How to propose an R & D plan ?

For important multi-national industrial companies, it is of great importance to maintain a high activity in Research and Developement (R & D). The company X is involved in the car industry sector. It has a high potential of internal R & D. It's annual budget is around 500 Millions French Francs (about $ 100 Millions). Before our study, no formal multicriteria evaluation procedure was in used to select the projects. In 1992, the R & D activities are divided into about 20 Technical Committees (TC) such as "Noise and Vibrations", "Security", "Comfort", ... These themes do not correspond strictly speaking to any formal hierarchical structure of the organization, but do have an existence of several years. They are transversal groups who are more specifically in charge of R & D in each respective area. Before our study, the annual R & D budget was established using the following classical procedure.

Each Technical Committee knows the budget he received the previous year. It has an idea of the global R & D new budget (same level or 5 % of increase, ...). Then each TC assumes that he would receive the same amount of increase. Therefore, he makes his choices at his level in order to propose the projects he prefers, so that the total cost should be compatible with the expected budget. The R & D plan is then just a collection of all these decentralized informal decisions. The possibility to compare projects of different TC is almost null, since the decision is already made at the level of each TC.

Two important factors were at the origin of a decision to implement a more formal multicriteria evaluation procedure.

First in 1991, it has been decided to reduce approximatively 20 % the level of expenses in R & D. Therefore, for the first time, the company had the opportunity to decide at a global level which projects should be rejected or postponed to the following years.

Second, a more significant change was promoted by the president of the company. Quickly speaking, it was necessary to promote quality, comfort, environmental considerations rather than pure mechanical performances.

P.M. Pardalos et al. (eds.), Advances in Multicriteria Analysis, 203–211.

Those two factors yield the necessity to have a more formal evaluation procedure of the R & D plan. To implement a formal evaluation procedure, it was decided to have a two level procedure.

The first level would consist in having a formal multicriteria evaluation procedure of the R & D projects whithin each TC. The aim was to have an overall ranking of the projects in greater number than the number of projects the company will be able to support. The comparison of the projects within each TC should be made on the basis of an internal policy of R & D in each field.

At a second level only, an R & D plan was established taking into account costs considerations and the strategic orientations of the company. We do not present in this paper the models wich were proposed at this level. They were based on a multiobjective (0/1) linear programming model. Since this model has not been implemented in the company, and because the present paper is devoted to an application of the UTA procedure, we do not present this multi-objective model in this paper.

2. Definition of the criteria

The definition of the criteria and operational means to evaluate the projects was the most important task of the study and also the most time consuming. The process took about 2 months. Five meetings of the study group, from 1 to 3 houres each, took place during this period. We suggested to study the list of criteria proposed by B. ROY and H. Le BOULANGER (1968). The six criteria they proposed (see also B. ROY, 1985 , page 348) were :

C1 - Annual cost of the project
C2 - Cost to be engaged in the future to implement the new research
C3 - Importance of the new markets achievable thanks to the results of the research
C4 - Strategic interest of the company for the project
C5 - Uncertainty in the results
C6 - Degree of success of previous phases of the same project

All theses criteria were supposed to have qualitative ordinal scales and a specific method was proposed by ROY and Le BOULANGER ("Méthodes des déclassements comparés"). From this initial list, our group study ended with the following list of criteria :

C1 - Expected progresses
C2 - Adequation of the research with the marketing departement demand
C3 - Degree of engagement of the research
C4 - Probability of success or failure
C5 - Adequation of the research with the basic technical orientations of the company
C6 - Timing of the research compared with the needs

Let us explain these criteria and see the associated ordinal scales.

C1 - Expected progresses

The way to evaluate a project on this criterion depends if the research is associated to a new product of the company of if is a research not directly associated to a new product (i.e. Information technology,...). In the first case, a scale was proposed combining subcriteria such as :
- importance of the products caracteristics
- progress normal or significant

A four grade scale combining these two attributes was chosen. A similar four grade scale was proposed for the research projects not directly associated to a new product.

C2 - Adequation of the research with the marketing department demandt

The following four-grade scale was defined :

e_4 - Project giving the company the capacity to achieve european norms for its products
e_3 - Project explicitly asked by the marketing department
e_2 - Normal evolution of the caracteristics of the products
e_1 - Normal evolution of the research in the field

C3 - Degree of engagment of the research

This criterion should promote projects which are either long terms (pluri-annual) projects almost achieved or short term projects yielding immediate results . A degree of achievment was defined using the following ratio :

$$d = 100 \,.\, \frac{\text{Previous cost + cost of the year}}{\text{Total cost of the project}}$$

It was decided to give an advantage to important projects which are almost finished. Therefore, the six grade criterion was established using the following table 1.

Cost d	0 - 500 KF	500 - 2 500 KF	> 2 500 KF
> 75	e_4	e_5	e_6
50 - 75	e_3	e_4	e_5
25 - 50	e_2	e_3	e_4
< 25	e_1	e_2	e_3

Table 1 . Aggregation of Cost and degree of achievement.(d in %)

<u>C4 - Probability of success or failure</u>

The following four-grade scale was defined :

e_4 - Low risk of failure and research easy to implement
e_3 - Risks limited to cost or delay increases
e_2 - Risks are possible not to achieve the desired level of the product caracteristic for which the research is proposed
e_1 - Research that can succeed but which remains difficult to implement

<u>C5 - Adequation of the research with the basic technical orientations of the company</u>

The four-grade criterion was defined :

e_4 - Research which can be done only by the company
e_3 - Research consistent with a relationship with another industrial partner
e_2 - Research neutral
e_1 - Research not well adapted to the technical orientation of the company

<u>C6 - Timing of the research</u>

This criterion appreciates the consistency between the date (year) of achievement of the research and the date (year) of the needs :

e_3 - Research is late - It is urgent to program it
e_2 - Research is in time
e_1 - Research can be postponed to next year

3. Which multicriteria method should be used ?

The company wanted to have an aggregation prodedure easy to understand and to explain. There was a strong preference towards simple additive rules such as a "<u>weighted sum of the criteria</u>". Although the scales were ordinal, the study group decided to assume an equivalence in the difference of each grade, and accepted therefore the cardinal structure necessary to use a simple additive rule.

There was also a wish to show that the aggregation procedure was as "objective" as possible. The weights should not represent any preference of a well defined decision-maker, but should be assessed as a joint collective view of the relative importance of the criteria.

Since there has been a relatively easy consensus on the choice of the criteria among the different TC, there was also a strong demand to converge if possible towards a common set of weights, which would represent therefore the rationalization of the R&D. policy making of the company. We decided therefore to use the model of a normalized weighted sum of the criteria :

$$u(a) = \sum_{i=1}^{n} w_i \quad (g_i(a) - g_{i*}) \; / \; (g_i^* - g_{i*})$$

With :

w_i = weight of the criterion i
$g_i(a)$ = evaluation of project a on criterion i
g_i^* = best preferred grade on criterion i (i.e. 4 for grade e_4)
g_{i*} = least preferred grade on criterion i (i.e. 1 for grade e_1)

In order to estimate a set of weights, it was decided to use a disaggregation procedure with each TC. Instead of asking the TC to weight the criteria, they were asked to give the following informations on the set of the projects they had selected for the previous R&D plan :

- give an evaluation to each project on each criterion using the scales associated to the criteria.
- give an overall judgment on the projects using the 3 following classes :

A - Project which is a top level project
B - Project that should be accepted in any case
C - Project that could be rejected or postponed to the following year

We must consider here that these classes do not allow us to assume that the overall preference has a weak-order structure. It has rather a partial order structure. More precisely, two projects put in class A are not necessarily equivalent, but a project in class A is preferred to a project in class B. Therefore, we used a special adaptation of the UTA procedure (see JACQUET-LAGREZE and SISKOS 1982) in order to estimate the weights. The adaptation of UTA is of a discriminant analysis type. The weights are estimated using a linear programming model which is presented in next section.

4. The UTA discriminant analysis model

The weights are estimated using UTA DIS, a discriminant analysis version of UTA Method. The linear programming formulation is the following :

Let G_k be the class k of project and K the number of classes. The K classes are ordered from best to worst. In our case study K = 3, G_1 = A, G_2 = B and G_3 = C.

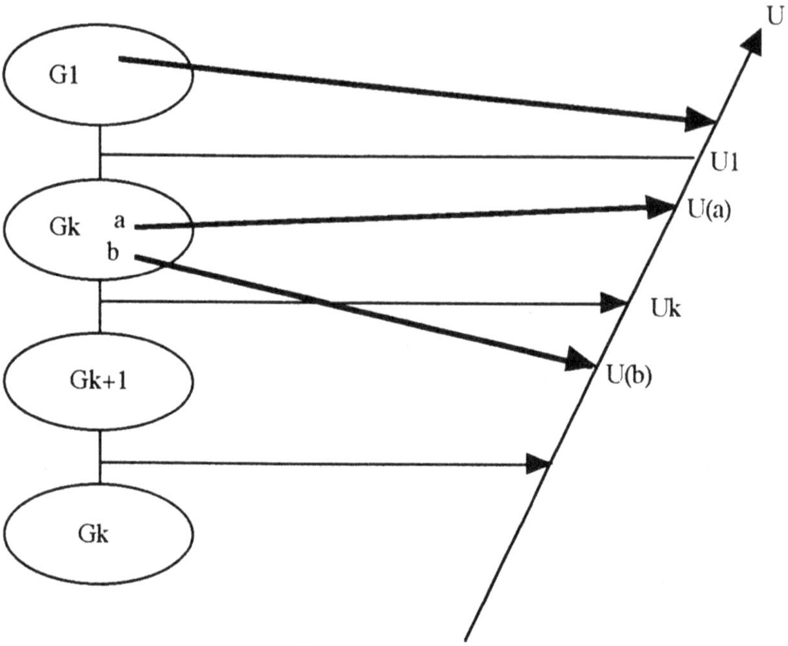

Fig. 1 UTA DIS : Mapping of classes in the utility scale

The unknown parameters of the model are the weights w_i of the criteria since we use a simple weighted sum of the criteria and the level of separation of thes classes on the overall scale (see Fig. 1):

u_1 - separation between class 1 and class 2

.

.

u_k - separation between class k and class k+1

Let us introduce for each project (alternative) two error variables allowing a project to be ill-classified (see Fig. 2). If the model is without error, we should have :

(1) $a \in G_k \quad \Leftrightarrow \quad u_{k+1} \leq u(a) \leq u_k$

Since the value of "a" can be greater than u_k or smaller than u_{k+1} , from (1) we derive the two inqualitу constraints, introducing two error variable $z^+(a)$ and $z^-(a)$:

(2 - 1)	u(a)	-	$z^+(a)$	$\leq$	u_k
(2 - 2)	u(a)	+	$z^-(a)$	$\geq$	u_{k+1}

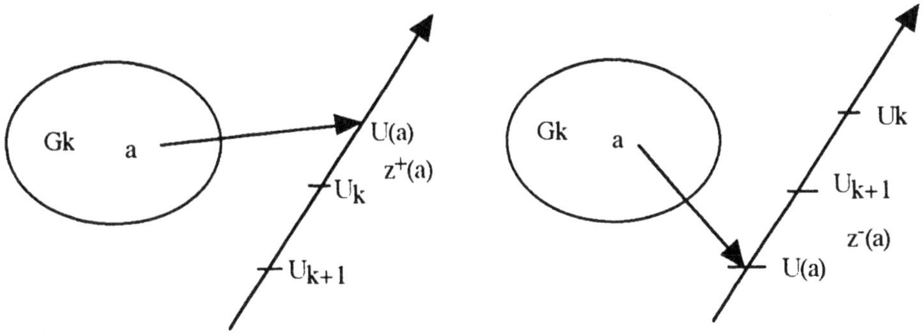

Fig. 2. Two cases for ill-classed projects

The decision variables of the model are :

w_i $i = 1, n$:	weight of the criteria
$z^+(a)$, $z^-(a)$	for all project $a \in A$
u_k $k = 1,$ $K - 1$	for all interclass level of utility

The linear Programming formulation is :

$$\text{Min} \quad \sum_{a \in A} \left(z^+(a) + z^-(a) \right)$$

$$\sum_i w_i \quad (g_i(a) - g_{i*}) / (g_i^* - g_{i*}) - z^+(a) - u_k \leq 0 \qquad \text{for all } a \in G_k$$

$$\sum_i w_i \quad (g_i(a) - g_{i*}) / (g_i^* - g_{i*}) + z^-(a) - u_{k+1} \geq 0 \qquad \text{for all } a \in G_k$$

$$\sum_i w_i = 1$$

$w_i \geq 0$ for all $i = 1,n$, $z^+(a)$, $z^-(a) \geq 0$ for all $a \in A$, $u_k \geq 0$ for all $k = 1, K\text{-}1$

5. - The process of estimating the weights

The weights were actually estimated using an interactive process based on the aggregation-disaggregation procedure (See JACQUET-LAGREZE, 1979). The decision-maker who represented the overall company's view was the responsible of the R & D plan. He had an idea of his a priori set of weights. But he did not want to communicate the set, feeling that his judgment was some what too subjective. So, we analysed first the set of weights estimated by UTA DIS approach with three different technical committees (TC1, TC2, TC3). The different results are given in table (2) :

	C1	C2	C3	C4	C5	C6
TC1 TC2 TC3	0,18 0,38 0,31	0,16 0,23 0,13	0,15 0,10 0,14	0,15 0,15 0	0,31 0,12 0,31	0,05 0,02 0,11
Average - 1	0,29	0,17	0,13	0,10	0,25	0,06
Average - 2	5	3	2	2	4	1
DM	3	4	1	1	2	2
Final	4	3	2	1	3	1

Table 2 - Different set of weights

The three first rows of the table are the results given by UTA-DIS analysis.

Then, the average set of weights of the three TC was computed (Average 1). In order to get integer weights easier to compare with the DM'set of weights we computed an approximate average set (Average 2) dividing each weight by the minimum value (0,06) and rounding the value to the nearest integer value. At that stage of the process, the DM revealed the weights he had previously written on a separate sheet of paper (row DM) - The final choice after some discussion with us was the row "Final" - Obviously, the final set shows some mutual influence between the disaggregation view of the TC obtained trough the UTADIS approach and the internal view of the DM.

6. - Conclusion

From a pure technical point of view, it would have been interresting to relax the assumption of linear additive model. The UTA discriminant disaggregation method can easily be extended to the case where marginal utility functions are piecewise linear. Probably, the adjustment would have been better. But our study group was not really willing to work with a more sophisticated non linear model.

The formal evaluation procedure using the six criteria and the set of weights obtained in this study, has been in use in the company during the following years. Before our study has begun, we wished to implement a more interactive procedure with a strong commitment of the Technical Committes. We were dreaming of a group decision meeting using computers and an interactive multicriteria aggregation model allowing the participants to change through the discussion their weigths, and eventually their own evaluation of the projects. But on the contrary, there was a strong demand of the company to have an aggregation rule as "neutral" and "objective" as possible. Therefore, the company appreciated the interactive process we used in order to estimate the weights using both the judgments of the TC on previous decisions and the a priori judgment of the DM (his own set of weights). From a more modest point of view, this process, which was real, illustrates in a certain way the aggregation-disaggregation process which can be used in multicriteria preference modelling.

REFERENCES

JACQUET-LAGREZE E. (1979) "De la logique d'agrégation de critères à une logique d'agrégation-désagrégation de préférences et de jugements", *Cahier de l('ISMEA*, Série Sciences de Gestion ,13,839-859

JACQUET-LAGREZE E. and SISKOS J. (1982) "Assessing a set of additive utility functions for multicriteria decision-making", *European Journal of Operational Research*, 10,2,151-164

Le BOULANGER and ROY B. (1968) "L'entreprise face à la sélection et à l'orientation des projets de recherche : la méthodologie en usage dans le groupe SEMA " - *Revue METRA*, vol VII, n°4, pp 641-669

ROY B. (1985) *Méthodologie Multicritère d'Aide à la Décision*, Economica, Paris.

Searching for a Financial Plan: A MOLP Multiple Reference Point Approach

João P. Costa and João N. Clímaco
Faculty of Economics, University of Coimbra
INESC Núcleo de Coimbra. Av Antero de Quental nº199
3000 Coimbra, PORTUGAL, Fax + 351 39 24692

Abstract

This paper presents a financial planning model and a multiple reference point approach (MRPA) to deal with it.

The model is formalised with a multiobjective linear programming (MOLP) approach considering a set of capital investment projects and a series of financing decisions over a planning period. Broadly speaking the financial decisions try to achieve a certain debt-equity ratio considering a firm's market value and the risk of financial distress. The key problem approached is the interaction of the financing and investment choices.

The MRPA enables oriented search for non-dominated solutions of a MOLP. Conceptually it is based on the quasi-satisfying rational framework for decision developed by Lewandovski and Wierzbicki [7]. It supports the decision maker (DM) learning about the non-dominated region to help him in the process of becoming aware of his aspirations and of their attainability. Based on the projection of the objective function gradients onto the non-dominated region the MRPA allows for the location of the reference points and their projections to travel in well defined and controlled directions, as well as the number of the reference points to change according to the exploratory needs of the DM.

Some tests carried out with the financial model are presented focusing the features that enable the choice of the search direction in a flexible and completely controlled (by the DM) manner. From the reported experiments the MRPA appears as being a useful approach to financial planning processes.

1. INTRODUCTION

According to Brealey and Myers [1] financial planning is a process of, among other actions, "analysing the interactions of the financing and investment choices" and "projecting the future consequences of present decisions in order to avoid surprises and understand the links between present and future decisions". Brealey and Myers [1] also say that there is no theory or model that leads straight to the optimal financial strategy.

P.M. Pardalos et al. (eds.), Advances in Multicriteria Analysis, 213–231.

Consequently financial planning proceeds by trial and error.

This trial and error process can be cumbersome due to the enormous quantity of computation and paperwork that it can generate. Like other planning processes (telephone network planning, for instance) the result, that is, the plan must be revised continuously by monitoring its performance against a rapidly changing environment. One can say that the plan it self is not the only benefit of the process but there is another, perhaps even more important: the knowledge acquisition by the people engaged in the planning process. The inside knowledge about how the environment usually changes and about the interactions between the decision instruments and data is of precious value.

In order to support such decision process it is necessary to provide managers with 'powerful' computational tools. However, when using these tools managers must be completely aware of what is happening to their plans. Moreover they must have the instruments to control the process and to find the plan that would satisfy them most. Besides that the tools must allow a constructive trial and error process (even a chaotic one if the manager feels comfortable with it), i. e., they must allow an easy learning of the non-obvious dependencies between the entities to deal with in the financial decision problem. If the tools are integrated in a decision support system (DSS) special efforts must be put in the design of the Human/machine interface to preserve the above features.

In this paper a multicriterion financial planning model and a multiple reference point approach (MRPA) [4] to deal with it are presented. The results of some experiments are also reported illustrating that the features outlined above (ii) can be achieved with the instruments that the MRPA can provide.

The model is formalised with a multiobjective linear programming (MOLP) approach therefore recognising explicitly that in a financial planning process there are several objectives in conflict to achieve a satisfying plan. This means that a financial plan ought always to be a compromise between opposing aspirations of managers. The model (see section 2) includes several features that are very useful in practice and enables high decision flexibility providing some instruments to achieve the aspirations of the decision maker. Nevertheless it is just one of the possible approaches and we do not claim that it is applicable to all financial planning problems. The key problem approached is the interaction of the financing and investment choices. More specifically, the problem is to decide how much to invest or borrow in the coming years to obtain a strategic plan. The model considers a set of capital investment projects (N is the number of projects) and a series of financing decisions over an H year planning period. Broadly speaking the financial decisions try to achieve a certain debt-equity ratio considering the firm's market value and the risk of financial distress. The model objective functions are the market value of the firm and one other for each planning year, which measures the capability of honouring promises to creditors at each planning year. It has therefore H+1 objective functions.

The literature (for instance [8]) extensively reports many interactive methods to deal with a MOLP approach, but among them and due to the characteristics of a financial planning process, the reference point methods appear as the most promising to deal with such problems. Usually people engaged in a financial planning process have to follow several recommendations that constitute the firm's policy or have to try to achieve some targets or aspirations proposed (or imposed) by other or themselves.

According to Lewandovski and Wierzbicki [7], one can say that "a decision maker (DM) behaves in a quasi-satisfying way if, aware of his objectives and aware of their attainability, he tries to reach his aspiration levels by maximising whenever the outcomes of admissible decisions fall below these levels. Meanwhile, if the aspiration levels are attainable, he can choose either to maximise (to reach efficient outcomes) or to forego maximisation for additional good reasons". Axiomatically one can say that if a DM has listed and modelled all relevant issues about a situation, then there exists no rational reason for him to be satisfied with solutions that are not non-dominated. The underlying problem is to find among all non-dominated solutions the one that best suits the aspirations of the DM. The DM aspiration levels define a reference point in the objective (or criterion) space that can be used to find a non-dominated solution of the MOLP problem. In the context of this communication DM should be understood as a complex entity that can be a single physical person or a set of persons the reason of which can achieve the same output to deal interactively with a financial planning problem (In any case it must be emphasized that this paper's focus is single DM problems.).

The MRPA [4] is presented in this paper as a means to help DMs in such a decision process. It is formalised in section 3 and, in section 4, the results of some experiments are reported. The goal of section 4 is to demonstrate that the MRPA when used or implemented in a DSS in a flexible way (not the way in which it is formalised in section 3) provides the above referred to features to people engaged in a planning process.

The core of the MRPA is composed of an achievement scalarizing function (see section 3.1), enabling to project a reference point, defined by the aspiration levels of the manager (DM), onto the non-dominated surface of the criterion space; and a gradient projected method (see section 3.2) allowing for a free 'travel' on the non-dominated frontier, with search directions under the control of the DM, in terms of the consequences to the objective function values.

Korhonen and Wallenius [6] developed an interactive MOLP method named 'Pareto Race', which was a pioneer decision support computational tool allowing for a free travel on the efficient frontier, enabling the DM to change the direction of motion in order to raise the value of an objective function. Although similar to Pareto Race in some points, the MRPA is more flexible concerning the DM's choice of the direction of search for the non-dominated solution preferred by the DM. The DM controls completely that direction and knows exactly what he is trading for when making a decision. This has been achieved by using directly a convex combination of the projections of the objective function gradients onto the non-dominated region as formalised in section 3 and illustrated in section 4.

The MRPA, based on the projection of the objective function gradients onto the non-dominated region, allows for the location of the reference points and their projections, to travel (i.e., to change by quasi continuous variations) in well defined and controlled directions, as well as the number of the reference points to change according to the exploratory needs of the DM. The most interesting features of this approach are those which enable:

(i) the choice of the search directions in a flexible and completely controlled (by the DM) manner, and

(ii) the variation of the number of reference points according to the exploratory needs of the DM.

Section 4 reports some tests carried out with the financial model focusing the features that enable the above point (i). Point (ii) is only briefly reported and its deep explanation can be considered out of the scope of this communication.

2. FINANCIAL PLANNING MODEL

In this section a financial planning model is presented. The model considers a set of capital investment projects (N is the number of projects) and a series of financing decisions over an H year planning period. Broadly speaking the financial decisions try to achieve a certain debt-equity ratio considering the firm's market value and the risk of financial distress. The model objective functions are the market value of the firm and one other for each planning year, which measures the capability of honouring promises to creditors at each planning year. It has therefore H+1 objective functions.

According to [1] the market value of the firm can be defined as the sum of the unlevered present value of the projects included in the investment plan, plus the tax savings implied by debt financing, minus the cost of issuing equity and minus the cost of holding cash. There is a fixed amount due to the market value of existing assets if they are all equity financed and to the debt already outstanding. The dependencies between the capital investment projects set and the firm's existing assets are assumed to be negligible by the firm. Therefore, that fixed amount is not relevant. We also assumed that the dividend payments are not relevant to the firm's market value.

The capability of honouring promises to creditors is measured by the expected earnings before interest and taxes minus the interest charge and should be maximised. Being a rough approximation to the non-linear definition of interest cover, i. e., the ratio of earnings before interest and taxes to the interest charges, this approach maintains its usefulness in most practice cases.

<u>The input data are the following</u>:

(i) Per year expected net cash flow per investment unit after taxes from the set of projects, F_{it}, where i=1, 2, ..., N and t= 0, 1, ..., H.

(ii) The maximum amount of investment per year on each project, M_i, where i=1, ..., N.

(iii) Per period contribution to earnings per investment unit before interest and taxes of projects, E_{it}, where i = 1, 2, ..., N and t = 0, 1, ..., H.

(iv) The rate of discount associated to each project (risk effects included), r_i, where i=1, 2, ..., N.

(v) The cost of issuing equity, c.

(vi) Interest rate on debt, r_b.

(vii) The expected return of the market, r.

(viii) Corporation marginal tax rate, T_c.

(ix) Target dividend payment ratio over earnings, TDE, and adjustment rate, AR.

(x) Minimum cash to hold as a percent of the earnings before interest and taxes, L_h.

(xi) The amount of cash at the beginning of the planning period, L_0.

<u>The set of instrumental (decision) variables are</u>:

x_i -> Amount of investment units on project i, where i=1, 2, ..., N.

y_t -> Amount of outstanding debt over year t, where t=1, 2, ..., H+1. $y_0 = 0$.

$y_t - y_{t-1} \leq 0$ -> Amount of matured debt obligations at the beginning of year t.

≥ 0 -> Amount of new debt obligations at the beginning of year t.

DIV_t -> Amount of dividend payments before taxes at the end of year t, where t=0,..., H.

SI_t -> Amount of equity issued at the end of year t, where t=0, 1, ..., H.

L_t -> Amount of cash to hold at year t, where t=1, 2, ..., H+1.

<u>The objective functions are</u>:

A) Firm's market value.

$$\max f_1 = \sum_{i=1}^{N} x_i \sum_{t=0}^{H} \frac{F_{it}}{(1+r_i)^t} + \sum_{t=0}^{H} \frac{T_c \cdot r_b \cdot y_{t+1}}{(1+r_b)^{t+1}} - \sum_{t=0}^{H} \frac{c.SI_t}{(1+r)^t} - \sum_{t=0}^{H} \frac{r.L_{t+1}}{(1+r)^{t+1}}$$

B) Capability of honouring promises to creditors.

$$\max f_{t+1} = \sum_{i=1}^{N} E_{it} . x_i - r_b . y_t \quad \text{where } t = 1, \cdots, H$$

<u>Constraints</u>:

A) Maximum amount of capital invested in each project.

For $F_{it} < 0$, i=1, ..., N, then

$F_{it} . x_{it} \leq M_i$

B) Cash flow for each year.

There are H+1 constraints, one for the beginning of each year, plus one for the end of the planning period. The cash flow from each project, plus the amount of new debt, minus the interest charges, plus the equity issued, minus the cost of issuing the equity, plus the cash of the previous year and minus the cash to be hold at the next year, equal zero.

$$\sum_{i=1}^{N} F_{it}.x_i + y_{t+1} - (1+r_b)y_t + (1-c)SI_t + L_t - L_{t+1} = 0 \quad \text{where } t = 0, \cdots, H.$$

C) Stabilisation of dividend payments around the ratio over earnings.

There are H+1 constraints. The first guarantees that the payment at the beginning of first year will be a weighted average over payments of the following years.

$$DIV_0 . \sum_{t=1}^{H} t - \sum_{t=1}^{H} (H+1-t).DIV_t \leq 0$$

The other H guarantee that the rise (or fall) of the payments from one year to the other equals the adjusted product of the dividend ratio over earnings for the year's total earnings minus the adjusted dividend payments of the previous year.

$$DIV_t - DIV_{t-1} = AR\ (\ TDE . \sum_{i=1}^{N} E_{it}.x_i - DIV_{t-1}) \quad \text{where } t = 1, \cdots, H$$

D) Debt outstanding.

There are H constraints guaranteeing that the objective functions associated to the capability of honouring compromises to creditors are equal or greater than zero.

$$\sum_{i=1}^{N} E_{it}.x_i - r_b.y_t \geq 0 \quad \text{where } t = 1, \cdots, H$$

E) Issuing equity.

There are H+1 constraints. One for the beginning of each planning year and another to the end of the planning period. These constraints guarantee that the amount of issued equity is equal or less than the amount of invested capital on new projects at the beginning of each year.

For each year t from 0 to H:

For each project i from 1 to N:

If $F_{it} > 0$ then $f_{it} = 0$ else $f_{it} = 1$.

$$\sum_{i=1}^{N} F_{it}.f_{it}.x_i + (1-c)SI_t \leq 0 \quad \text{where } t = 0, \cdots, H.$$

F) Holding cash.

In order to be prepared for unexpected expenses the firm's management considers necessary to hold at least a percentage, L_h, of the year's expected earnings before interest and taxes. So there are H constraints guaranteeing this management caution.

$$L_h \sum_{i=1}^{N} E_{it} . x_i - L_t \leq 0 \quad \text{where } t = 1, \cdots, H.$$

Besides all of these there are several underlying assumptions. Broadly speaking:

a) The model is deterministic. The uncertainty inherent to the forecast of the economical parameters should be accounted for exogenously. The model only supports the uncertainty generated by the managers' dynamic attitude towards the conflicting goals of the decision problem

b) The projects are economically independent both mutually and from existing operations.

c) Besides taxes, transaction costs and soft bounds to credit, the market is considered to be perfect.

d) All the cash flow can be delayed to the beginning of each year.

e) The dividend payments don't have effects on the firm's market value. The goal of the management is to stabilise these payments over a dividend payments-earnings ratio target.

3. MULTIPLE REFERENCE POINT APPROACH - MRPA

The first step of the MRPA is to project a reference point defined by the aspiration levels of the DM onto the non-dominated region following the direction defined by some scaling units that are also defined by the DM. Section 3.1. presents an achievement scalarizing program that can be used in this first step.

If this projection of the reference point is to be moved on the non-dominated region, a direction of movement must be found. To let the DM choose this direction it is necessary to apply a gradient projected method, the two phases of which are formalised in sections 3.2 and 3.3 respectively. In order to be decision-support effective this gradient projected method must be applied in a flexible way (see section 4). Sections 3.2 and 3.3 present a rigid algorithm providing the method foundations but not very useful in practice due to its lack of flexibility.

In the first phase, the gradients of the objective functions of the MOLP are projected onto the non-dominated region, defining p directions (p = number of criteria), each of them being as close to the gradients of each objective function as possible. There are as many sets of p directions as active constraints in the non-dominated solution obtained through the projection of the reference point. Each of these sets of directions enables to travel on the non-dominated sub-region (hyper-plane) corresponding to each of the active constraints. Note that if one of the active constraints is dominated then the gradient projections onto it will only enable to travel on the intersection of this constraint with one of the non-dominated active constraints at that point.

For each of these sets of directions, a reference point can be defined if the DM wishes to explore the non-dominated sub-region associated with the corresponding active constraint. Therefore, it is on the intersection of two or more active constraints that new reference points are created according to the exploratory needs of the DM, enabling him to explore new non-dominated sub-regions located after the last known frontier.

The inner product of each objective function normalised gradient with their normalised projections onto the non-dominated region, on each set, define a value, h^u_i (i =1, ..., p ; u = 1, ..., k ; k = number of active constraints) which is a measure of the maximum proximity of criterion i that one can achieve when travelling on the non-dominated sub-region associated with the set of directions, u. The process of choosing a travelling direction, which is a convex combination of the objective function gradient projections onto a chosen non-dominated sub-region (constraint) is the second phase of the method (formalised in section 3.3.). Knowing the h_i value (the travelling maximum possible proximity of direction i), the DM defines iteratively for each of the objective functions the minimum proximity that is acceptable. On each iteration the h_i coefficients are recalculated for the other criteria. For example, if at the first iteration the DM sets the minimum acceptable proximity of criterion j as g_j ($g_j \leq h_j$), the projections of the other criteria onto the non-dominated sub-region to be explored are recalculated. Each of these projections (defining d'_i) is as close to the gradients of the corresponding objective function as possible, taking into account that their proximity to the criterion j lies in the interval $[g_j , h_j]$. At the second iteration the process is repeated for other criterion different from j with the d'_i coefficients.

Consider now that the solution is travelling on a non-dominated sub-region following a chosen direction. What happens if the DM intends to change the travelling direction? If the solution arrives to an intersection of two or more constraints then the DM must change the direction of travelling, and must fulfil the first and second phases of the method. If the last change of direction had occurred when the solution was on an intersection of two or more constraints, then the DM must fulfil the first and second phases of the method, if he wants to change the travelling direction. In the other cases the DM only needs to fulfil the second phase of the method.

The needed calculations for travelling on a non-dominated sub-region, following a chosen direction, are presented in section 3.4. Note that the computations are done on the decision space.

3.1. Achievement scalarizing program.

A multiobjective linear program (MOLP), see Steuer [8], can be written as:

$$\max \{ \mathbf{c}^1 \mathbf{x} = q_1 \}$$
$$\max \{ \mathbf{c}^2 \mathbf{x} = q_2 \}$$
$$\ldots$$
$$\max \{ \mathbf{c}^p \mathbf{x} = q_p \}$$
$$\text{s. t. } \mathbf{x} \in S$$
$$\text{or "max" } \{ C\,\mathbf{x} = \mathbf{q} \mid \mathbf{x} \in S \}$$

where p is the number of objectives; $\mathbf{c}^i \in \mathbf{R}^n$, is the gradient (vector of objective function coefficients) of the ith objective function; q_i is the criterion value (objective function value) of the ith objective; S is the feasible region; $\mathbf{x} \in \mathbf{R}^n$, is the decision vector; "max" indicates that the purpose is to maximise all objectives simultaneously; C is the p x n criterion matrix (matrix of objective function coefficients) whose rows are the gradients $\mathbf{c}^i$ of the p objective functions; and $\mathbf{q} \in \mathbf{R}^p$ is the criterion vector (objective function vector).

Feasible region in criterion space:

Whereas S denotes the feasible region in the decision space, Q denotes the feasible region in the criterion space:

$Q = \{ \mathbf{q} \in \mathbf{R}^p \mid \mathbf{q} = C\,\mathbf{x}, \mathbf{x} \in S \}$

in other words, Q is the set of images of all points in S.

Achievement scalarizing function:

Let a given $\mathbf{q}^* \in \mathbf{R}^p$ be denoted a reference point in the criterion space. The reference point may or may not be feasible (i. e. be a member of Q). Then, an achievement scalarizing function $s(\mathbf{q}^*,\mathbf{q})$ is one that projects the reference point $\mathbf{q}^*$ onto the set of non-dominated criterion vectors. The projected point on the set of non-dominated criterion vectors is obtained by solving the achievement program:

$\max \{ s(\mathbf{q}^*,\mathbf{q}) \}$

s.t. $\mathbf{q} \in Q$

The following achievement function is used (a simplified version of the one found in [7]):

$$s(\mathbf{q}^*,\mathbf{q}) = \min_{1 \le i \le p} \frac{q_i - q_i^*}{s_i} + \frac{\varepsilon}{p} \sum_{i=1}^{p} \frac{q_i - q_i^*}{s_i}$$

Where the parameter ε should be kept positive, even if very small. If this parameter was equal to zero, then the maximal points of the above function could correspond to weakly non-dominated solutions. The s_i is a scaling unit for the variable i.

With this function, the achievement scalarizing program is formulated as a single criterion program:

$$\max \left\{ \alpha + \frac{\varepsilon}{p} \sum_{i=1}^{p} \frac{q_i - q_i^*}{s_i} \right\}$$

$$\text{s.t.}: \quad c^i x = q_i \qquad 1 \le i \le p$$

$$\alpha \le \frac{q_i - q_i^*}{s_i} \qquad 1 \le i \le p$$

$$x \in S$$

$$\alpha \in \mathbf{R}$$

3.2. Projecting the objective function gradients.

First it is necessary to find the active constraints corresponding to the non-dominated solution computed with the achievement scalarizing program (section 3.1) or corresponding to the current non-dominated solution if it is a result of a previous movement on the non-dominated region. Consider that the lines, a_{i*} , i = 1, ..., k, of the matrix A (which defines S : $A\,x = b$) are associated with the k active constraints.

To project the gradients of the objective functions of the MOLP onto the non-dominated region, computing directions that are as close to the gradients of each objective function as possible, it is necessary to solve the following linear programs:

For each u = 1, ..., k.

For each t = 1, ..., p.

$\max\ \mathbf{c}^t\,\mathbf{x}$

s.t. $a_{i*}\ \mathbf{x}\ [\le, \ge, =]\ 0$, $i = 1, \ldots, k\ ;\ i \ne u$. (The type of each constraint is the same as the corresponding constraint in the original problem.)

$a_{u*}\,\mathbf{x} = 0$

$\forall\ \mathbf{x}$

The solution of a program like this is the vector zero, or the program is unbounded. If the program is unbounded, the projection of an objective function gradient is the unbounded direction that leads to the infinite solution and it can be obtained from the Simplex tableau. Note that the constraints of non-negativity are also considered as potentially active and it is not necessary to alter the above referred to procedure if the non-dominated solution is degenerated.

For each u a set of p directions is computed (even if some of them are the vector zero) and so there is a set of directions associated with each of the active constraints. For each of these sets, a convex combination can be obtained and used as the exploring direction of the non-dominated sub-region associated with that active constraint.

3.3. Choosing a direction of search.

To explore a particular non-dominated sub-region, let us say u, the DM must choose a convex combination of the set of objective function gradient projections associated with that sub-region. The process of choosing this travel direction is iterative. At each iteration the DM defines his minimum acceptable travelling proximity for each objective function of the MOLP. The following algorithm is only a formalised and rigorous way of presenting this process. Through extended experimentation it has revealed no practice use due to its lack of flexibility to effectively support a trial and error decision process.

The process can be described like this:

1. Calculate the inner product of each objective function normalised gradient with its normalised projections onto the non-dominated sub-region, u, (computed in the way presented in section 3.2.): h_i ; $i = 1, ..., p$. The h_i value is a measure of the maximum proximity of criterion i, that one can achieve when travelling on the non-dominated sub-region that the DM wants to explore.

2. Without loss of generality, let $i = 1$. (For simplicity the DM will begin by the first criterion.)

3. Knowing h_j , $j = 1, ..., p$, the DM defines g_i ($g_i \leq h_i$), the minimum acceptable travel proximity of criterion i.

4. Calculate the following linear programs:

For each $t = i+1, ..., p$

$$\max \mathbf{c}^t \mathbf{x}$$

$$\text{s.t.} \quad \mathbf{c}^r \mathbf{x} \geq g_r \qquad r = 1, ..., i.$$

$$\mathbf{x} - \sum_{i=1}^{p} \mu_i \mathbf{d}_i \;, \mu_i \geq 0 \;, \mathbf{d}_i = \text{the } i^{th} \text{ criterion gradient projection.}$$

$$| \mathbf{x} |_{L_1} = 1$$

$$\forall \mathbf{x}$$

5. Recalculate the inner product of each objective function normalised gradient with its normalised projections onto the non-dominated sub-region, u, recalculated at point 4: h_j , $j = i+1, ..., p$.

6. Without loss of generality let $i = i + 1$. If $i < p$ go to point 3, else stop.

3.4. Travelling on a non-dominated sub-region.

If the direction of movement was determined as presented in sections 3.2 and 3.3, the following formulas can be applied to travel on a non-dominated sub-region.

Suppose that the non-dominated solution to be moved following the direction $\mathbf{d}$, is $\mathbf{x}^1$.

$\mathbf{x}^2$, defined by $\mathbf{x}^2 = \mathbf{x}^1 + \Delta\ \mathbf{d}$, is yet a non-dominated solution if $\Delta \in [\Delta_1 , \Delta_2]$, where:

$$\Delta_1 = \max_i \left(-\frac{x_i^1}{d_i} \right) \text{ if } d_i > 0 \text{ and } \Delta_2 = \min_j \left(-\frac{x_j^1}{d_j} \right) \text{ if } d_j < 0 .$$

4. TESTS AND RESULTS

In this section some results of an application of the multiple reference point method to the financial planning model of section 2, are presented. The presentation of these results enables to illustrate some decision support features of a computational approach to the method the core of which is formalised in section 3. These features were derived through extended experimentation with the financial model and focusing on the decision problem of choosing a direction of search for other non-dominated solutions. The input data (based in [9]) used for tests, for a six year planning period, is presented in table I.

	Y. o		Y. 1		Y. 2		Y. 3		Y. 4		Y. 5		Y. 6		M	r
P.	F	E	F	E	F	E	F	E	F	E	F	E	F	E	10^3	%
1	-10		2	1.5	2	1.5	2	1.5	2	1.5	2	1.5	2	1.5	30	10.4
2			-10	*	2	1.5	2	1.5	2	1.5	2	1.5	2	1.5	30	10.4
3					-10	*	2	1.5	2	1.5	2	1.5	2	1.5	30	10.4
4							-10	*	2	1.5	2	1.5	2	1.5	30	10.4
5	-50		10	7	10	7	10	7	10	7	10	7	10	7	70	9.2
6	-200		22	2	22	2	22	2	22	2	22	2	22	2	30	7.2
7	-100		50	10	50	15	50	25							50	12.8
8							-100		50	10	50	15	50	25	50	12.8
9			-50		10	2	11	4	12	6	13	8	14	10	70	13.2

c	2%	r	10%	TDE	40%	L_h	10%
r_b	12%	T_c	50%	AR	50%	L_0	$10 000

Table I - Input data. The currency unit is the dollar.

The first step is to find a non-dominated solution according to the manager's aspiration levels and scaling units (see section 3.1). The manager's aspiration level for the firm's market value is 200 000 dollars, a value that he knows to be impossible to achieve. In what concerns the capability of honouring promises to creditors the manager wants to be in the middle, i. e., he wants that the expected earnings double the interest charges for each year. Knowing the expected maximum earnings the management fixed the following aspiration levels (one for each planning year, respectively): 9 800, 14 700, 20 850, 23 400 and 27 300 dollars. Due to the risk always involved in forecasting, management considers that its aspiration levels for the last years are more important than the others, more precisely, it considers that they are twice more important for the third and fourth years and three more times important for the fifth and sixth years than for the first and second.

The firm's market value is considered to be four times more important than the aspiration levels for the first and second years. The resulting non-dominated solution after solving the achievement scalarizing program, see section 3.1, is presented in table II. The objective function (o.f.) values are depicted in the bar graph of fig. 1.

Decision variable values							
x_1	3 000.0	x_2	3 000.0	x_3	3 000.0	x_4	3 000.0
x_5	1 400.0	x_6	26.4	x_7	500.0	x_8	500.0
x_9	1 214.3	y_1	161 273.2	y_2	239 845.7	y_3	248 422.3
y_4	301 542.6	y_5	275 248.1	y_6	246 257.3	y_7	208 778.2
SI_0	0	SI_1	0	SI_2	0	SI_3	0
DIV_0	14 509.2	DIV_1	13 140.2	DIV_2	12 326.4	DIV_3	14 305.2
DIV_4	15 180.4	DIV_5	16 603.7	DIV_6	18 801.1	L_1	1 935.3
L_2	2 878.1	L_3	4 071.0	L_4	4 013.9	L_5	4 506.8
L_6	5 249.6	L_7	0				
Objective function values							
f_1	143 689.0	f_2	0	f_3	0	f_4	10 899.4
f_5	3 953.8	f_6	11 287.0	f_7	22 194.6		

Table II - The initial non-dominated solution of the financial planning problem.

The manager is pleased in what concerns the firm's market value (It is the maximum.), but he is strongly unsatisfied with the capability of honouring promises at the first and second years of the planning period (o.f. two and three), also disliking the value for the fourth year (o.f. five). Note that these results were somewhat expected. The manager's aspiration level to the firm's market value can not be achieved and the other aspiration levels are in the middle of the possibilities. On the other hand, the scaling units were defined by the DM giving more importance to the firm's market value. Therefore it was expected from the firm's market value to reach its maximum. Analysing table II one can see that only projects 9 and 6 were not completely fulfilled. Project 9 has a negative net present value, NPV, nevertheless the plan recommends to invest in it. This is due to the NPV of the tax benefits and to the new horizons that this investment opens to the firm (to have more cash at a future year, for instance). An equilibrium has been achieved. In what concerns project 6 the drawback is its low amount of earnings before interest and taxes.

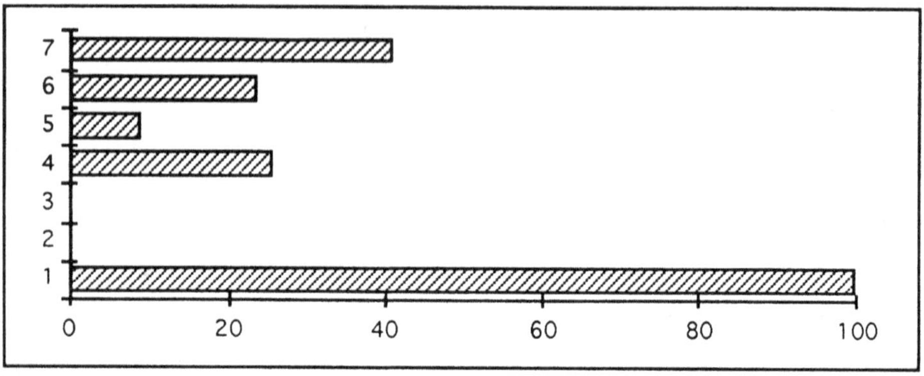

Fig. 1 - Objective function values as a percentage of their maximum.

Now the MRPA allows the manager to change the non-dominated solution in order to achieve a higher degree of satisfaction. The non-dominated solution can travel over distinct non-dominated hyper-faces: one for each present active constraint the sign of which is not an equal and one another corresponding to the equal constraints. The MRPA demands the DM to explore each one of these hyper-faces (non-dominated sub-regions). It demands to project the criteria's gradients over each one of these non-dominated sub-regions and then expects the DM to choose a direction of search in each one. There are ways of disregarding this MRPA's exhaustive demand and selecting only the most promising non-dominated sub-regions to explore, according to the physical characteristics of the financial planning problem and to the manager's preferences. To discuss these ways is out of the scope of this study. The most promising sub-regions to explore originate different paths of search and due to the correlation between each path and a reference point movement (in terms of aspiration levels or scaling units learning evolution) originate different reference points. Considering that one non-dominated sub-region has been selected as the most promising to explore the interactive process of choosing a direction of search will be deeply illustrated.

Projecting the objective function gradients over the non-dominated sub-region corresponding to the active constraints the sign of which is an equal, the manager obtains a measure of the maximum possible proximity of each criterion that one can achieve when travelling on this non-dominated sub-region. These results are presented to the manager in a non-normalized fashion providing him with a better comprehension of what will happen to the objective function values. These results are presented in table III, interaction A. For instance, the maximum possible travel proximity to objective function two is 4.9 cents per each travelling unit. Note that each one of these proximity values can correspond to different direction of search.

	Measures of the maximum possible proximity of each criterion.						
Interactions	f_1	f_2	f_3	f_4	f_5	f_6	f_7
A	0.0	4.9	2.1	3.1	5.5	8.1	10.7
B	-3.3	**3.4**	1.4	3.1	5.5	8.1	10.7
B 1	-2.4	**2.5**	1.6	3.1	5.5	8.1	10.7
C	-4.3	**2.5**	**1.0**	3.0	5.1	7.2	9.4
D	**-4.3**	**2.5**	**1.0**	2.1	3.0	4.1	5.2
E	**-4.3**	**2.5**	**1.0**	2.0	3.0	4.1	**4.0**
F	**-4.3**	**2.5**	**1.0**	2.0	3.0	**3.9**	**4.0**
G	**-4.3**	2.5	**1.0**	2.0	**2.4**	**3.9**	**4.0**
H	**-4.3**	2.5	**1.0**	**2.0**	**3.0**	**4.1**	5.2

Table III - Results of the interactions to choose a direction of search.

Decision variable values changes per unit.							
x_1	0	x_2	0	x_3	0	x_4	0
x_5	0	x_6	- 1.156 E-3	x_7	0	x_8	0
x_9	+ 3.915 E-3	y_1	- 2.276 E-1	y_2	- 3.735 E-2	y_3	- 5.295 E-2
y_4	- 7.258 E-2	y_5	- 9.603 E-2	y_6	- 1.234 E-1	y_7	- 1.595 E-1
SI_0	0	SI_1	+ 5.933 E-3	SI_2	0	SI_3	0
DIV_0	+ 3.784 E-3	DIV_1	+ 1.430 E-3	DIV_2	+ 1.818 E-3	DIV_3	+3.579 E-3
DIV_4	+ 6.025 E-3	DIV_5	+ 8.813 E-3	DIV_6	+ 1.177 E-2	L_1	- 2.312 E-4
L_2	+ 5.518 E-4	L_3	+ 1.335 E-3	L_4	+ 2.118 E-3	L_5	+2.901 E-3
L_6	+ 3.684 E-3	L_7	0				
Objective function values changes per unit.							
f_1	- 4.300 E-2	f_2	+ 2.500 E-2	f_3	+ 1.000 E-2	f_4	+ 1.970 E-2
f_5	+ 2.988 E-2	f_6	+ 4.053 E-2	f_7	+ 5.164 E-2		

Table IV - The direction of search, both in the spaces of decision and of the criteria.

Due to his dissatisfaction, the manager chooses to set a minimum raise to o.f. two (capability of honouring promises at the first year) of 70% of its maximum rise. The measures of proximity to the other criteria are recalculated taking into consideration this constraint. The manager is therefore confronted with the results presented in table III, interaction B. Analysing the results the manager feels that to have a fall of at least 3.3 cents on the firm's market value for a minimum raise of 3.4 cents on the o.f. two is not comfortable, considering that he wants also to raise other objective functions. So he lows the minimum raise of o.f. two to only 50% (2.5 cents) of its possible maximum.

Recalculating the proximity measures he is confronted with the results of table III, interaction B1. He then fixes the minimum raise for o.f. three to 60% (1 cent) of its actual maximum. With this additional constraint he is trading a market value fall of at least 4.25 cents for a raise of at least 2.5 cents and 1 cent of the capability of honouring promise at years 1 and 2 respectively. See table III, interaction C. Note that on this stage the manager could try to lower again the minimum raise constraint of o.f. two, but for raising it up, it would first be necessary to set it free, and then to fix it again (for knowing its possible maximum).

Decision variable values							
x_1	3 000.0	x_2	3 000.0	x_3	3 000.0	x_4	3 000.0
x_5	1 400.0	x_6	0.0	x_7	500.0	x_8	500.0
x_9	1 303.7	y_1	156 073.6	y_2	238 992.6	y_3	247 213.0
y_4	299 542.6	y_5	273 055.0	y_6	243 438.7	y_7	205 135.2
SI_0	0	SI_1	135.5	SI_2	0	SI_3	0
DIV_0	14 145.6	DIV_1	13 172.8	DIV_2	12 367.9	DIV_3	14 387.0
DIV_4	15 318.0	DIV_5	16 805.0	DIV_6	19 069.9	L_1	1 930.0
L_2	2 890.7	L_3	4 101.5	L_4	4 062.2	L_5	4 573.0
L_6	5 333.7	L_7	0				
Objective function values							
f_1	142 707.0	f_2	570.9	f_3	228.4	f_4	11 349.3
f_5	4 636.4	f_6	12 212.6	f_7	23 373.9		

Table V - The non-dominated solution of the financial planning problem at the end of the direction of search.

The manager doesn't want to lower any more the value of o.f. one. So he fixes its fall to at most 4.3 cents. Analysing the computations at this point, see table III interaction D, the manager fixes the raising value for o.f. seven to at least 4 cents. He then realises that he made a mistake, because among the remaining o.f.s the value he most dislikes is the value of o.f. five. But he also notices, see table III interaction E, that the constraint associated to the rise of o.f. seven value didn't have effects on the maximum rise of o.f. five, i.e., there is no conflict in the non-dominated sub-region under analysis between the o.f. seven and the o.f.s five and six. Encouraged with these results the manager iteratively fixes at 3.9 cents and 2.4 cents the raise of the values of o.f.s six and five. He obtains the results presented at table III interactions F and G.

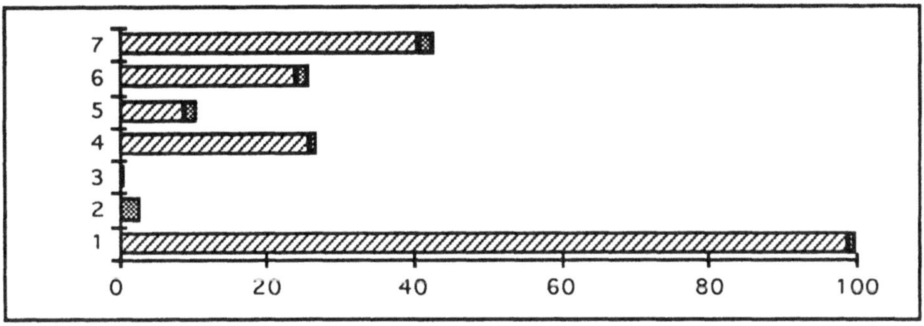

Fig. 2 - Objective function values and changes as a percentage of their maximum.

Table III, interaction H, presents the real objective function values' changes, that is not the 'at most' or 'at least' values, when travelling in the non-dominated sub-region following the direction chosen by the DM. These results are also presented at table IV with the direction of search. This direction can be considered both in the spaces of decision and of the criteria.

Analysing table IV the manager knows exactly what he is doing when considering this direction of search. He can see for instance that for lowering the firm's debt over the planning years (y variables) he must let the firm's market value fall and he must invest more in project 9 and less in project 6. Therefore this approach privileges the earnings over the NPV. He can also see the amounts of trade between the economical entities that the travelling process implies.

The o.f. values and their maximum changes are depicted at fig. 2. These maximum changes are calculated as presented at section 3.4. The manager can travel according to this direction until reaching an edge of the non-dominated region (which was the present case) or until he decides to choose another direction of search. The 'excursion' (path) according to the direction of table IV is a short one due to the low amount of the reported investment in project 6 (26.4 x 200 dollars, see table II). Therefore the changes on the objective function values can not be very big. Nevertheless the manager is travelling in a good direction to achieve his satisfying plan. Reaching the non-dominated plan presented in table V the manager could stop or choose another (or other) non-dominated sub-region(s) to explore.

Note that there is a direction of movement of the manager's aspiration levels corresponding to this direction of search, considering the same scaling units, or thinking reversibly, there is a way of changing the scaling units corresponding to this direction of search, considering the same aspiration levels.

5. CONCLUDING REMARKS

From the above reported experiments the MRPA appears as being a useful approach to financial planning processes giving effective decision support due to the following:

i) It can fit well with some financial planning methods by trying to achieve a set of targets and then by allowing a process of incremental changes (travelling on the non-dominated frontier) to the obtained plan.

ii) It enables a constructive process of reaching the financial plan providing managers with a set of flexible control instruments the concept of which are easy to capture due to its closeness to the field's intuition of managers. Managers don't need to know anything about MOLP theory to choose a direction of search. The decisions are all taken in a financial context.

iii) It supports a learning process about the financial problem by presenting to mangers the trades linking the economic entities involved in the process of changing plans to reach a satisfying one.

Note that it is not easy to describe such an interactive process due to the difficulty of capture the learning experience of using directly a computational decision support tool. A more conclusive remark about the usefulness of the MRPA will need more experience on supporting financial decision processes.

REFERENCES

1. Brealey, A., C. Myers. *Principles of Corporate Finance*. Mc Graw-Hill, 1991.
2. Clímaco, J. N., J. P. Costa, C. H. Antunes, Maria J. Alves. *Parallel processing in MOLP method base development - A discussion using two case studies*. Research Report, 1992.
3. Costa, J. P., J. N. Clímaco. *Dynamic multiobjective optimisation - Parallel aspects of achievement functions*. In "Parallel Computing' 91", Elsevier Science Publishers, D. J. Evans, G.R. Joubert and H. Lidell (Editors), pp 593-600, 1992.
4. Costa, J. P., J. N. Clímaco. *A Multiple Reference Point Parallel Approach in MCDM*. Multiple Criteria Decision Making. Expand and enrich the Domain of Thinking and Application. Springer Verlag. G.-H. Tzeng, H.-F. Wang, U.-P. Wen and P. L. Yu, (Editors), pp 255-263, 1994.
5. Jacquet-Lagreze, E. *A projected gradient method for linear programming*. Cahier du Lamsade no 79, 1987.
6. Korhonen, P., J. Wallenius. *A Pareto Race*. Naval Research Logistics 35, pp 615-623, 1983.
7. Lewandovski, A. and A. Wierzbicki ; *Aspiration based decision analysis and support.*

Part I: theoretical and methodological backgrounds. IIASA - 88 - 03, 1988.

8. Steuer, R. *Multiple criteria optimisation: theory, computation and application.* John Wiley & Sons, New York, 1986.
9. Spronk, J. *Interactive Multiple Goal Programming. Applications to Financial Planning.* International Series in MS/OR. Martinus Nijhoff Publishing, 1981.

SCHEDULING SURGERIES FOR PATIENTS REQUIRING POST-OPERATIVE INTENSIVE CARE: A MULTIPLE OBJECTIVE INTEGER PROGRAMMING APPROACH

Harold P. Benson

College of Business Administration
University of Florida, Gainesville, Florida, 32611, U.S.A.
and
Health Services Research and Development, VA Medical Center,
Gainesville, Florida, 32602, U.S.A.

Abstract

One of the more challenging tasks at the Veteran's Administration Hospital in Gainesville is that of scheduling patients for surgery. In this paper, we present a linear multiple objective integer programming model for scheduling surgeries at the VA which require post-operative intensive care. The model can also be used to explore the tradeoffs in the number of surgeries of this type that can be achieved. We also indicate how the model is used, and we provide some sample results which demonstrate the practicality and validity of the model.

P.M. Pardalos et al. (eds.), Advances in Multicriteria Analysis, 233–247.

1. Introduction

Perhaps one of the more challenging tasks at the Veteran's Administration Hospital in Gainesville is that of scheduling patients for surgery. For many types of scheduled surgery, there is usually a long list of patients waiting to undergo that surgery. For surgeries that require post-operative care in the intensive care unit, the scheduling task is even more difficult. This is because the numbers of nurses and beds available in the intensive care unit are limited. Presently, the Chief of Surgery schedules surgeries requiring post-operative intensive care without the aid of any formalized procedures for maximizing the number of surgeries provided.

To potentially aid in providing more efficient schedules and to help analyze various aspects of the surgery scheduling system, a linear multiple objective integer programming model has been developed and tested. Preliminary experience with this model indicates that it is a practical and potentially quite useful model for scheduling surgeries which require post-operative intensive care and for exploring various tradeoffs in the number of surgeries of this type that can be achieved.

The purpose of this paper is to describe the nature and use of the linear multiple objective integer programming model for scheduling surgeries requiring post-operative intensive care. In Section 2, the nature of the surgery scheduling problem addressed by the model is briefly described. In Section 3, as a preliminary to the presentation of the model, an explanation is given of what a linear multiple objective integer program is, what its advantages are, and how it can be solved. In Section 4, this model is presented and explained. In Section 5 we indicate how the model is used, and, in Section 6, some sample results are given which demonstrate the practicality and validity of the model. Some concluding remarks are given in Section 7.

2. Description of the Surgery Scheduling Problem

In this Section we will briefly describe the nature of the surgery scheduling problem that is addressed by the model. In order to do so, we must first describe the system used for providing surgical services in the VA hospital.

Currently, a wide variety of types of scheduled surgery are performed in the hospital. These surgeries are performed in a total of seven operating rooms, five days per week. For purposes of assigning surgeries to operating rooms, 13 categories of surgery have been defined at the hospital. Eight of these categories together utilize over 78 per cent of the operating room time which is available, and each of these surgeries is called a "major user" of operating room time. The remaining time is used by surgery types which are each referred to as a "minor user" (see Table 2.1).

Depending upon the type of surgery that has been performed, a patient is either placed in the surgical intensive care unit (SICU) or in the ward for post-operative recovery. The SICU is a specialized care center receiving patients who generally have undergone more serious operations than those who proceed directly to the ward following surgery. It is equipped with 9 beds, one of which is kept for emergencies only and another of which is a spare bed. Three shifts of nurses are used to staff

Surgery Categories	
Major Users	Minor Users
General	Dental
Vascular	Podiatry
Cardio-Thoracic	Gynecology
Neurosurgery	Oncology
Plastic Surgery	Radiation Therapy
Urology	
Ear, Nose, and Throat	
Eye	

Table 2.1 Surgery Categories

the SICU 24 hours per day. One head nurse, three licensed practicing nurses, 20 registered nurses, and two technicians work in the SICU. The length of time a patient remains in the SICU depends mainly upon the type of surgery he has undergone. The type of care he requires each day of his stay in the SICU depends upon both the type of surgery he has undergone and the number of days that have passed since his surgery. Following his stay in the SICU, a patient is discharged to the ward for further recovery. Although many types of surgeries require recovery in the SICU, about 90 per cent of the beds in the SICU are occupied at any given time by patients who have undergone one of four types of surgery. These types are general, vascular, cardio-thoracic, and neurosurgery.

For many types of scheduled surgery, there is usually a long list of patients waiting to undergo the surgery. Usually the hospital is able to provide adequate operating room staff to meet the daily surgery demand at the hospital. However, there are a number of other resources whose supplies appear to be more critical in limiting the rate at which surgeries can be performed. Among these resources are the number of operating rooms, the number of beds available in the SICU, and the number of nurses available in the SICU. If more of these resources were available, it would probably be possible to service patients awaiting surgery at a more rapid rate. However, since these resources are basically fixed at certain levels, an alternate means must be found if the number of surgeries performed per day is to increase.

One means of influencing the ability of this system to provide surgical services is by carefully planning weekly schedules of surgeries that are to take place. If schedules could be found which efficiently use the available resources, then the rate at which surgeries are performed might be increased. One step towards this goal is to determine upper limits on the ability of this system to provide various surgical services. With a knowledge of these upper limits, a great deal of insight could be gained. For instance, one would be able to answer questions such as the following:

(i) What is the maximum number of surgeries that could be performed in one week if the system operates as expected?

(ii) If one would like to guarantee that at least four cardiothoracic surgeries are performed in a given week, what numbers of surgeries of other types would be possible if the system performs as expected?

(iii) If vascular surgeries are to receive more preference than neurosurgeries in a given week, and the system performs as expected, then how many of these and other surgeries could be performed in that week?

The linear multiple objective integer programming model that is proposed in this paper is intended as a management aid for answering questions such as those posed above. Since many of the critical resources of the system are associated with the SICU, the model deals only with those four categories of surgeries that account for 90 per cent of the bed use in the SICU. To use the model, one must first choose a positive weight for each of these four surgery categories. These weights correspond to the relative importance one would like to associate with each category. The model then finds the number of operations in each of the four categories that maximizes the weighted sum of the number of operations that can be performed in one week, assuming that the system operates as expected. It also specifies a daily schedule of surgeries which achieves this surgery plan. By varying the choice of the four weights, the model can be used to generate a number of alternate surgery plans. In this way, one can explore the tradeoffs that can potentially be achieved among the levels of the four surgeries. This tradeoff information would then allow one to answer questions such as those posed above for these four surgery types.

3. A Brief Overview of Linear Multiple Objective Integer Programming

Integer programming is one of the most widely-used methods for solving practical problems involving the allocation of scarce resources. Various methods for solving integer programs have been formulated over the past 40 years, and integer programming has been used extensively to solve problems arising in both the private and public sectors. Applications of integer programming have been in areas such as manufacturing, transportation, water resources, facility location, energy planning, investment, education, and hospital administration, to name a few (see [1–3] for a detailed discussion of integer programming and its applications).

One of the disadvantages of integer programming is that it requires that a single objective function be chosen for maximization or minimization. In many cases, more than one objective function exists. This is especially true in public sector problems such as location, transportation, education, and health care problems. To help analyze such situations, various approaches are possible. These include *goal programming* [4], *multiattribute utility theory* [5], and *multiple objective linear* and *nonlinear programming* [6]. The model proposed in this paper is a linear multiple objective integer programming model. To aid in our presentation of this model, let us briefly explain what a linear multiple objective integer program is, what its advantages are, and how it can be solved.

Let $x = (x_1, x_2, \ldots, x_n)$ be a vector of integer decision variables. The decision vector x specifies levels of activities to be undertaken. Let $f_j(x)$ be a linear function of x whose value specifies, for each choice of x, the level attained by x in objective number j, $j = 1, 2, \ldots, p$, where $p \geq 2$. In addition, let $g_i(x)$ be a linear function of x whose value represents, for each x, the amount of resource i consumed, $i = 1, 2, \ldots, m$. Assume, for each $i = 1, 2, \ldots, m$, that b_i units of resource i are available. Then, the ***linear multiple objective integer programming problem*** (P)

$$\text{"maximize"} \ [f_1(x), f_2(x), \ldots, f_p(x)]$$

subject to

$$g_i(x) \leq b_i, \ i = 1, 2, \ldots, m \tag{1}$$

$$x \geq 0, \ x \text{ integer}, \tag{2}$$

seeks to find the set of all vectors x^0 that are efficient in the sense of the following definition.

DEFINITION. A vector x^0 is said to be an *efficient* solution of (P) when x^0 satisfies (1) and (2), and there is no vector x satisfying (1) and (2) such that $f_j(x) \geq f_j(x^0)$, $j = 1, 2, \ldots, p$, with $f_k(x) > f_k(x^0)$ for at least one k in $\{1, 2, \ldots, p\}$.

We will refer to any vector x which satisfies (1) and (2) as a ***feasible solution*** for (P). The linear functions $f_j(x)$, $j = 1, 2, \ldots, p$, will be called the ***objective functions*** for (P). Loosely speaking, then, an efficient solution of (P) is a feasible solution for (P) for which no other feasible solution exists which achieves at least as large a value in every objective function and a strictly larger value in at least one of the objective functions. Such solutions are also called ***nondominated*** or ***Pareto-optimal*** solutions of (P). The linear multiple objective integer programming problem (P) seeks to generate the set of all of these efficient solutions.

The linear multiple objective integer programming approach can be used in any situation in which the decision variables are integer and one seeks to maximize or minimize more than one linear objective function subject to linear constraints. It is especially attractive when it is difficult or impossible to combine the objective functions into a single objective function. There are at least two reasons for this.

First, provided that an increase in the value of any objective function is not detrimental, the efficient solutions are the only feasible solutions for (P) that need to be considered in making a decision as to which feasible solution to implement. To see this, suppose hypothetically that we are able to specify a utility function $U[f_1(x), f_2(x), \ldots, f_p(x)]$ of the objective functions which gives the "value" of any decision vector x. In general, since the objective functions are noncomparable, such a utility function does not exist. However, assuming that it does exist, if an increase in the value of any objective function is not detrimental, then U must be nondecreasing in its arguments $f_1(x), f_2(x), \ldots, f_p(x)$. Because of this characteristic of U, it can be easily shown that as long as *some* vector maximizes $U[f_1(x), f_2(x), \ldots, f_p(x)]$ over all feasible solutions x, there will exist an efficient plan which does so also. This implies that as long as increases in the objective functions are not detrimental to our goals, the efficient solutions are the only solutions that need to be considered.

A second advantage of the linear multiple objective integer programming approach is that it reveals the tradeoffs that can be achieved among the objective functions. After all of the efficient solutions have been generated, this tradeoff information can be obtained by simply examining their objective function values. For instance, consider an efficient solution x^0 and its value $f_1(x^0)$ in the first objective function. If some other efficient solution x^1 satisfies $f_1(x^1) > f_1(x^0)$, then, by definition of an efficient solution, $f_j(x^1)$ will be less than $f_j(x^0)$ for one or more objective functions f_j, $j \neq 1$. Therefore, if, in order to increase the value of f_1, x^1 is chosen instead of x^0, certain other objective function values will decrease by certain observable amounts. By examining all of the efficient solutions x which satisfy $f_1(x) > f_1(x^0)$, then, one is able to see how much the value of f_1 could be increased at the expense of various decreases in other objective functions if solutions alternate to x^0 are chosen. Such tradeoff information can be very useful in decision-making situations with multiple objective functions.

Various algorithms have been proposed for generating the set of all efficient solutions of (P) (see [7–9], for instance). However, the computational burden of generating this set grows rapidly as the number of variables n or the number of constraints m in (P) increases. Therefore, methods for generating subsets of the set of all efficient solutions are often used. We have used one such method to generate some efficient solutions for our surgery scheduling model. To close this section, we will briefly explain this method.

Let $w_1, w_2, \ldots, w_p$ represent positive weights. Consider the problem (P_w)

$$\text{maximize} \sum_{j=1}^{p} w_j f_j(x)$$

subject to

$$g_i(x) \le b_i, \quad i = 1, 2, \ldots, m$$
$$x \ge 0, \; x \text{ integer.}$$

It is easily shown that x^0 is an efficient solution of (P) if x^0 is an optimal solution of problem (P_w) for some set of positive weights $w_1^0, w_2^0, \ldots, w_p^0$. Therefore, to generate various efficient solutions of (P), one can choose various sets of positive weights $w_1, w_2, \ldots, w_p$ and solve (P_w) for each set of weights chosen. This is the method we used for the surgery scheduling model. For any given efficient solution x^0, one can interpret the associated weights $w_1^0, w_2^0, \ldots, w_p^0$ as describing the relative importance that the solution x^0 assigns to each objective function.

4. The Multiple Objective Surgery Scheduling Model

The surgery scheduling model is a linear multiple objective integer programming model which seeks to find efficient weekly surgery schedules for the four major operating room users that account for about 90 per cent of the bed use in the SICU. These users are general, vascular, cardiothoracic, and neurosurgery. For each set of four positive weights chosen for these four surgery categories, an efficient schedule

Category	Number	Subcategory	Number
General	1	Colon Resection	1
		AP Resection	2
		Pancreatic	3
		Biliary	4
		Gall Bladder	5
		Stomach	6
		Thyroid	7
Vascular	2	Aneurysmorraphy	8
		Aortobifemoral	9
		Femoral Popliteal or Tibial Bypass	10
		Axillobifemoral	11
		Carotidendarterectemy	12
		Thrombectomy	13
Thoracic	3	Coronary Artery Bypass	14
		Aortic/Mitral Valve	15
		Combined (AVR & CABG)	16
		Thoracotomy	17
Neurosurgery	4	Craniotomy	18
		Carotid	19
		MCA-STA Bypass	20
		Spinal Cord Injury	21

Table 4.1. Surgery Subcategories

can be found. In this way, representative efficient schedules can be generated which reveal the tradeoffs that can potentially be achieved among the levels of the four surgery types. This tradeoff information then allows one to answer questions such as those posed in Section 2 of this paper.

To present the model, we will first describe its variables and parameters. Next, the model itself will be given. Last, we will discuss the meaning of each of the objective functions and constraints of the model.

4.1 Variables and Parameters

Let the four categories of surgery, general, vascular, thoracic, and neurosurgery, be numbered 1, 2, 3, and 4, respectively. In the model, these four categories are subcategorized to yield a total of 21 subcategories of surgery. These subcategories are given in Table 4.1. Let the subcategories be numbered as in Table 4.1, and let the days of the week Monday through Friday be numbered one through five, respectively. Then the three main types of variables in the model can be described as follows.

(1) For each $i = 1, 2, \ldots, 21$ and $j = 1, 2, 3, 4, 5$, x_{ij} represents the number of surgeries of subcategory i to schedule on day j.

(2) For each $i = 1, 2, \ldots, 21$, y_i represents the number of surgeries of subcategory i to schedule in the week.

(3) For each $k = 1, 2, 3, 4$, z_k represents the number of surgeries of category k to schedule in the week.

In addition to these variables, there are eight types of parameters in the model. These are as follows.

(1) For each $j = 1, 2, 3, 4, 5$, s_j represents the operating room time available on day j, in days, for performing surgeries of types one through four.

(2) For each $i = 1, 2, \ldots, 21$, t_i represents the average time required, in days, to perform one operation of subcategory i. The time t_i includes any required setup time and time for administering anesthesia.

(3) For each $i = 1, 2, \ldots, 21$, r_i represents the average number of recovery days in the SICU required for a patient who has undergone surgery of subcategory i. The day of surgery is designated as day 0 and the day after surgery as recovery day number one. On day r_i the patient requires no bed or nurse care in the SICU and is discharged from the SICU to the ward. For each i, $1 \le r_i \le 3$.

(4) For each $i = 1, 2, \ldots, 21$ and each $h = 0, 1, 2, \ldots, r_i$, m_{ih} represents the average number of minutes of nurse care required in the SICU on day h for a patient who has undergone surgery of subcategory i and is in day h of his recovery in the SICU.

(5) For each $i = 1, 2, \ldots, 21$, L_i represents a lower bound on the desired number of surgeries of subcategory i to be performed during the week.

(6) For each $i = 1, 2, \ldots, 21$, U_i represents an upper bound on the desired number of surgeries of subcategory i to be performed during the week.

(7) The parameter a represents the number of minutes of nurse care available in the SICU in a 24-hour period for patients who have undergone surgeries of any of the subcategories $1, 2, \ldots, 21$.

(8) The parameter b represents the average number of beds available each weekday in the SICU for patients who have undergone surgeries of any of the subcategories $1, 2, \ldots, 21$.

4.2 Statement of the Model

The linear multiple objective integer programming model (S) for determining efficient weekly schedules for the four categories of surgery described previously can now be stated

$$\text{“maximize”} \quad [z_1, z_2, z_3, z_4] \tag{3}$$

subject to

$$\sum_{i=1}^{21} t_i x_{ij} \leq s_j, \; j = 1,2,3,4,5 \tag{4}$$

$$\sum_{i=1}^{21} x_{i1} \leq b \tag{5}$$

$$\sum_{\{i|r_i \geq 2\}} x_{i1} + \sum_{i=1}^{21} x_{i2} \leq b \tag{6}$$

$$\sum_{\{i|r_i \geq 3\}} x_{i1} + \sum_{\{i|r_i \geq 2\}} x_{i2} + \sum_{i=1}^{21} x_{i3} \leq b \tag{7}$$

$$\sum_{\{i|r_i \geq 3\}} x_{i2} + \sum_{\{i|r_i \geq 2\}} x_{i3} + \sum_{i=1}^{21} x_{i4} \leq b \tag{8}$$

$$\sum_{\{i|r_i \geq 3\}} x_{i3} + \sum_{\{i|r_i \geq 2\}} x_{i4} + \sum_{i=1}^{21} x_{i5} \leq b \tag{9}$$

$$\sum_{i=1}^{21} m_{i0} x_{i1} \leq a \tag{10}$$

$$\sum_{\{i|r_i \geq 2\}} m_{i1} x_{i1} + \sum_{i=1}^{21} m_{i0} x_{i2} \leq a \tag{11}$$

$$\sum_{\{i|r_i \geq 3\}} m_{i2} x_{i1} + \sum_{\{i|r_i \geq 2\}} m_{i1} x_{i2} + \sum_{i=1}^{21} m_{i0} x_{i3} \leq a \tag{12}$$

$$\sum_{\{i|r_i \geq 3\}} m_{i2} x_{i2} + \sum_{\{i|r_i \geq 2\}} m_{i1} x_{i3} + \sum_{i=1}^{21} m_{i0} x_{i4} \leq a \tag{13}$$

$$\sum_{\{i|r_i \geq 3\}} m_{i2} x_{i3} + \sum_{\{i|r_i \geq 2\}} m_{i1} x_{i4} + \sum_{i=1}^{21} m_{i0} x_{i5} \leq a \tag{14}$$

$$\sum_{\{i|r_i \geq 3\}} m_{i2} x_{i4} + \sum_{\{i|r_i \geq 2\}} m_{i1} x_{i5} \leq a \tag{15}$$

$$\sum_{\{i|r_i \geq 3\}} m_{i2} x_{i5} \leq a \tag{16}$$

$$\sum_{j=1}^{5} x_{ij} - y_i = 0, \; i = 1,2,\ldots,21 \tag{17}$$

$$\sum_{\{i|i\in I_k\}} y_i - z_k = 0, \; k = 1, 2, 3, 4 \tag{18}$$

$$L_i \leq y_i \leq U_i, \; i = 1, 2, \ldots, 21 \tag{19}$$

$$z_k \geq 0, \; k = 1, 2, 3, 4 \tag{20}$$

$$x_{ij} \geq 0 \text{ and integer}, \; i = 1, 2, \ldots, 21; \; j = 1, 2, 3, 4, 5. \tag{21}$$

In the statement of model (S), expressions of the form $\sum_{\{i|r_i \geq q\}}$ denote that the indicated summation is to be performed over all indices i in $\{1, 2, \ldots, 21\}$ that satisfy the condition $r_i \geq q$. In (18), for each k, the summation is performed over all indices i in $\{1, 2, \ldots, 21\}$ which also lie in I_k, where I_k contains the index numbers of the subcategories of surgeries in category k (for instance, $I_1 = \{1, 2, \ldots, 7\}$ and $I_2 = \{8, 9, \ldots, 13\}$).

4.3 Interpretation of the Model

The objective functions z_1, z_2, z_3, and z_4 of the multiple objective surgery scheduling model (S) are given in (3). The model seeks to find schedules which are efficient as measured by the numbers of surgeries of categories one through four (general, vascular, thoracic, neurosurgery) performed per week.

The constraints in (4) require, for each weekday, that the total operation time devoted to performing surgeries of the four categories not exceed the operating room time available. Since, for each $i = 1, 2, \ldots, 21$, t_i represents the average time required to perform an operation of subcategory i, we assume in (4) that any operation proceeds at the expected rate.

Constraints (5) through (9) require, for Monday through Friday, respectively, that the number of beds occupied in the SICU by patients who have undergone surgeries of categories one through four not exceed the number of beds available for such patients in the SICU. For instance, any patient operated upon on Wednesday, any patient operated upon on Tuesday whose average recovery time is two days or more, and any patient operated upon on Monday whose average recovery time is three days or more will require a bed in the SICU on Wednesday. The left-hand side of constraint (7) specifies the total number of such patients. Thus, constraint (7) requires that the total number of patients who require a bed in the SICU on Wednesday who have undergone surgeries of categories one through four not exceed the number of beds expected to be available for them on that day. A similar explanation holds for constraints (5), (6), (8), and (9). No constraints of this type are required for Saturday or Sunday since those two constraints hold automatically when (9) holds. Since average recovery times and average numbers of beds available are used in constraints (5) through (9), we assume in (5) through (9) that recovery period lengths are as expected.

Constraints (10) through (16) require, for each of Monday through Sunday, respectively, that the average time required by nurses in the SICU to care for the patients who have undergone surgeries of types one through four not exceed the total time available in that day for such care. The explanation and meaning of each of the individual constraints (10) through (16) are similar to those of constraints (5) through (9). However, since the minutes of nurse care required by a patient in a day depends partially upon the number of days since his operation, the constraints for Saturday and Sunday [(15) and (16)] may not be implied by the others in this case. In (10) through (16), we assume, on any given recovery day, that any patient requires the average amount of nurse care on that day of all patients who have undergone the same surgery. Thus, in (10) through (16), we assume that nurse care requirements are as expected.

The constraints in (17) and (18) simply define the variables $y_1, y_2, \ldots, y_{21}$ and z_1, z_2, z_3, and z_4 in terms of the variables x_{ij}, $i = 1, 2, \ldots, 21$; $j = 1, 2, 3, 4, 5$. The constraints in (19) require, for each $i = 1, 2, \ldots, 21$, that the number of surgeries of subcategory i scheduled in the week lie between the chosen lower and upper bounds. These bounds are included to guarantee that certain demand levels for surgeries are met and to model other technical and practical considerations. The constraints of types (20) and (21) are nonnegativity and integer conditions. Since, for each $i = 1, 2, \ldots, 21$, we can assume that $L_i \geq 0$, in any feasible solution y_i is automatically nonnegative for each i by (19). Therefore, no nonnegativity requirements need to be given for these variables. Also, from (17), (18), and (21), in any feasible solution, y_i and z_k are automatically integers for each i and k.

5. Application of the Multiple Objective Surgery Scheduling Model

To use the linear multiple objective integer programming model (S) to find efficient weekly schedules at the VA hospital for the four major categories of surgery considered in the model. two preliminary activities must be undertaken.

First, values for the eight types of parameters described in Section 4.1 must be gathered which correspond to the situation in the VA hospital. Values for these data that we have used in our sample runs (see Section 6) are presented in Tables 5.1 and 5.2. In Table 5.2, for each $i = 1, 2, \ldots, 21$, since average values were not available, the values of t_i, r_i, and m_{ih}, $h = 0, 1, 2, 3$, are minimum, rather than average, values. Also in Table 5.2, for each i, the values of L_i and U_i given are those that most commonly occur in any given week. For any particular week, alternate values for these parameters may be chosen which are more appropriate. Alternate values would be more appropriate when alternate demands or alternate technical or practical situations occur. All of the data in Tables 5.1 and 5.2 were gathered by the Chief of Surgery at the hospital and his assistants.

The second preliminary activity that must be undertaken in order to find efficient surgery schedules for the VA hospital using model (S) is to obtain a computerized procedure for solving the problem (P_w) obtained in this case for each set of positive weights w chosen (see Section 3 for the general form of problem (P_w)). For each set of weights w, problem (P_w) is a linear integer program. There are various commercially-available computer codes for solving linear integer programming problems. The one that we used to solve problem (P_w) for each set of weights w chosen in our sample runs is called the Mathematical Programming System Extended, or MPSX [10]. The main program which accomplished this is as follows.

s_1	=	4 days
s_2	=	4 days
s_3	=	3 days
s_4	=	4 days
s_5	=	2 days
b	=	3 beds
a	=	3780 nurse-minutes

Table 5.1 **Operating Room Time, SICU Beds, and SICU Nurses-Minutes Available at the VA (See Section 4.1 for explanation of parameters)**

Surgery Subcategory i	t_i (Days)	r_i (Days)	m_{i0}	m_{i1} (Minutes)	m_{i2}	m_{i3}	L_i	U_i
1	.2500	1	735	0	0	0	0	1
2	.2916	2	903	861	0	0	0	1
3	.3542	3	840	840	840	0	0	1
4	.2916	3	840	840	840	0	0	2
5	.2083	1	630	0	0	0	0	1
6	.2708	1	966	0	0	0	0	1
7	.2813	1	630	0	0	0	0	1
8	.2708	3	1302	903	819	0	0	2
9	.2916	3	966	966	966	0	0	2
10	.2708	1	861	0	0	0	0	2
11	.2708	1	861	0	0	0	0	1
12	.2292	1	651	0	0	0	0	3
13	.2083	1	651	0	0	0	0	3
14	.2083	2	1407	1113	0	0	2	5
15	.2813	3	1407	1113	966	0	0	2
16	.2916	3	1407	1113	966	0	0	1
17	.2708	2	1323	840	0	0	2	5
18	.2916	2	672	672	0	0	1	3
19	.2500	1	630	0	0	0	0	1
20	.3125	1	735	0	0	0	0	1
21	0.0000	3	1260	1260	1260	0	0	1

Table 5.2 **Operating Times, Recovery Times, SICU Nurse-Care-Minutes, and Bounds for the 21 Surgery Subcategories at the VA (see Section 4.1 for explanation of parameters)**

Weight Set Number	Weight Values			
	w_1	w_2	w_3	w_4
1	1.0	1.0	1.0	1.0
2	1.0	1.0	2.0	1.0
3	1.0	1.5	2.0	1.0
4	0.7	1.2	1.5	1.0
5	1.0	2.0	3.0	1.0
6	1.0	2.0	3.0	1.5

Table 6.1. Weight Sets Used in Sample Runs

```
PROGRAM
INITIALZ
TITLE('EFFNT')
MOVE(XDATA, 'SICU')
MOVE(XPBNAME, 'VA')
MOVE(XRHS, 'EF')
MOVE(XOBJ, 'OPNS')
CONVERT('SUMMARY')
SETUP('MAX', 'BOUND', 'NOP')
CRASH
PRIMAL
OPTIMIX('MATRIX',0.,3,0,0)
EXIT
PEND
```

6. Sample Results

In order to test the practicality and validity of the surgery scheduling model (S), we have generated various efficient weekly schedules using the data given in Section 5. Based upon these sample results, we have found the model to be both practical and valid. To see how we reached this conclusion, we will briefly summarize the results of our sample runs of the model.

To test model (S), we chose six sets of weights w_1, w_2, w_3, and w_4 and solved problem (P_w) for each set of weights using MPSX. These six sets of weights are given in Table 6.1. For each set of weights, we obtained an efficient mix of surgeries and at least one weekly schedule for achieving this mix (when (P_w) had multiple optimal solutions, we obtained more than one such schedule). On the average, problem (P_w) was solved by MPSX in about 1.36 seconds of central processing unit time on an Amdahl 470 V/6 II computer. Although the solution time may vary with other choices of weights or bounds L_i and U_i, $i = 1, 2, \ldots, 21$, this small average solution time indicates that model (S) is quite practical if problem (P_w) is used to generate efficient schedules.

The validity of the model was evident from the reasonableness of the efficient weekly surgery schedules that were generated. To illustrate the nature of these schedules, we will summarize the efficient schedule generated by weight set number 1.

The efficient schedule found using weight set number 1 calls for three general surgeries ($z_1 = 3$), four vascular surgeries ($z_2 = 4$), four thoracic surgeries ($z_3 = 4$), and two neurosurgeries ($z_4 = 2$) to be performed in one week. Since this is an efficient schedule, there is no other schedule with values for z_1, z_2, z_3, and z_4 satisfying $z_1 \geq 3$, $z_2 \geq 4$, $z_3 \geq 4$, and $z_4 \geq 2$ with strict inequality holding at least once. This particular efficient schedule calls for a total of 13 surgeries of the four major category types considered. This is quite reasonable since currently about 9 to 10 such surgeries are performed per week.

To achieve these four surgery levels, the efficient schedule calls for the following plan.

Day of Week	Types and Number of Surgeries to Perform
Monday	1 colon resection 2 carotidendarterectemies
Tuesday	1 gall bladder 1 thyroid 1 carotid
Wednesday	1 thrombectomy 2 thoracotomies
Thursday	1 carotidendarterectemy
Friday	2 coronary artery bypasses 1 craniotomy

Using this plan, each of the three beds available in the SICU for patients undergoing surgeries of the types considered is filled each of Monday through Friday. This also is quite reasonable, since intuitively the number of these beds available seems to be the most critical resource in the surgery scheduling system considered by the model.

7. Concluding Remarks

We have presented and explained the use of a linear multiple objective integer programming model for scheduling surgeries and exploring the tradeoffs among surgeries which require post-operative intensive care. The sample results indicate that the model is both practical and of potential use for surgery scheduling in the VA. To use the model, one merely needs to choose lower and upper bounds for the 21 individual surgeries and weights for the four objective functions. The lower and upper bounds may vary from week to week as surgeries are scheduled due to changes in demands for surgeries or changes in other practical or technical areas.

The weights chosen correspond to the importance being placed on the four major surgical services considered in the model.

References

[1] R. S. Garfinkel and G. L. Nemhauser, *Integer Programming*, Wiley, New York, N.Y., 1972.

[2] H. M. Salkin, *Integer Programming*, Addison-Wesley, Reading, Massachusetts, 1975.

[3] G. L. Nemhauser and L. A. Wolsey, *Integer and Combinatorial Optimization*, Wiley-Interscience, New York, N.Y., 1988.

[4] J. P. Ignizio, *Goal Programming and Extensions*, D. C. Heath, Lexington, Massachusetts, 1976.

[5] R. L. Keeney and H. Raiffa, *Decisions with Multiple Objectives: Preferences and Value Tradeoffs*, Wiley, New York, N.Y., 1976.

[6] H. W. Kuhn and A. W. Tucker, "Nonlinear Programming," in: *Proceedings of the Second Berkeley Symposium on Mathematical Statistics and Probability*, J. Neyman, ed., University of California Press, Berkeley, California, 1950, pp. 481–492.

[7] G. R. Bitran, "Linear Multiple Objective Programs with Zero-One Variables," *Math. Programming*, **13** (1977) 121–139.

[8] R. Banker, M. Guignard, and S. Gupta, "An Algorithm for Generating Efficient Solutions to the Multiple Objective Integer Programming Problem," Applied Research Center, The Wharton School, University of Pennsylvania, Philadelphia, PA., December, 1978.

[9] D. Klein and E. Hannan, "An Algorithm for the Multiple Objective Integer Programming Problem," *European J. Operation. Res.*, 9 (1982) 378–385.

[10] International Business Machines, *Mathematical Programming System Extended (MPSX), Mixed Integer Programming (MIP) Program Description*, IBM, White Plains, N.Y., 1973, 2nd edition.

AUTHOR INDEX

Nonconvex Optimization and Its Applications

1. D.-Z. Du and J. Sun (eds.): *Advances in Optimization and Approximation.* 1994. ISBN 0-7923-2785-3
2. R. Horst and P.M. Pardalos (eds.): *Handbook of Global Optimization.* 1995 ISBN 0-7923-3120-6
3. R. Horst, P.M. Pardalos and N.V. Thoai: *Introduction to Global Optimization* 1995 ISBN 0-7923-3556-2; Pb 0-7923-3557-0
4. D.-Z. Du and P.M. Pardalos (eds.): *Minimax and Applications.* 1995 ISBN 0-7923-3615-1
5. P.M. Pardalos, Y. Siskos and C. Zopounidis (eds.): *Advances in Multicriteria Analysis.* 1995 ISBN 0-7923-3671-2

KLUWER ACADEMIC PUBLISHERS – DORDRECHT / BOSTON / LONDON

www.ingramcontent.com/pod-product-compliance
Lightning Source LLC
LaVergne TN
LVHW011752200926
848484LV00003B/684
* 9 7 8 0 7 9 2 3 3 6 7 1 6 *